Design, Empathy, Interpretation

Design Thinking, Design Theory

Ken Friedman and Erik Stolterman, editors

Design, Empathy, Interpretation

Toward Interpretive Design Research

Ilpo Koskinen

The MIT Press
Cambridge, Massachusetts
London, England

The MIT Press would like to thank the anonymous peer reviewers who provided comments on drafts of this book. The generous work of academic experts is essential for establishing the authority and quality of our publications. We acknowledge with gratitude the contributions of these otherwise uncredited readers.

This book was set in Stone Serif and Stone Sans by Westchester Publishing Services. Printed and bound in the United States of America.

Library of Congress Cataloging-in-Publication Data

Names: Koskinen, Ilpo, author.
Title: Design, empathy, interpretation : toward interpretive design research / Ilpo Koskinen.
Description: Cambridge, Massachusetts : The MIT Press, [2023] | Series: Design thinking, design theory | Includes bibliographical references and index.
Identifiers: LCCN 2022054175 (print) | LCCN 2022054176 (ebook) | ISBN 9780262546928 (paperback) | ISBN 9780262376884 (epub) | ISBN 9780262376877 (pdf)
Subjects: LCSH: Design—Research.
Classification: LCC NK1520 .K68 2023 (print) | LCC NK1520 (ebook) | DDC 744.072—dc23/eng/20230302
LC record available at https://lccn.loc.gov/2022054175
LC ebook record available at https://lccn.loc.gov/2022054176

10 9 8 7 6 5 4 3 2 1

Contents

As professions go, design is relatively young. The practice of design predates professions. In fact, the practice of design—making things to serve a useful goal, making tools—predates the human race. Making tools is one of the attributes that made us human in the first place.

Design, in the most generic sense of the word, began over 2.5 million years ago when *Homo habilis* manufactured the first tools. Human beings were designing well before we began to walk upright. Four hundred thousand years ago, we began to manufacture spears. By forty thousand years ago, we had moved up to specialized tools.

Urban design and architecture came along ten thousand years ago in Mesopotamia. Interior architecture and furniture design probably emerged with them. It was another five thousand years before graphic design and typography got their start in Sumeria with the development of cuneiform. After that, things picked up speed.

All goods and services are designed. The urge to design—to consider a situation, imagine a better situation, and act to create that improved situation—goes back to our prehuman ancestors. Making tools helped us to become what we are: design helped to make us human.

Today, the word "design" means many things. The common factor linking them is service, and designers are engaged in a service profession in which the results of their work meet human needs.

Design is first of all a process. The word "design" entered the English language in the 1500s as a verb, with the first written citation of the verb dating to the year 1548. *Merriam-Webster's Collegiate Dictionary* defines the verb "design" as "to conceive and plan out in the mind; to have as a specific purpose; to devise for a specific function or end." Related to these is the act of drawing, with an emphasis on the nature of the drawing as a plan or map, as well as "to draw plans for; to create, fashion, execute or construct according to plan."

Half a century later, the word began to be used as a noun, with the first cited use of the noun occurring in 1588. *Merriam-Webster's* defines the noun as "a particular purpose held in view by an individual or group; deliberate, purposive planning; a mental project or scheme in which means to an end are laid down." Here, too, purpose and planning toward desired outcomes are central. Among these are "a preliminary sketch or outline showing the main features of something to be executed; an underlying scheme that governs functioning, developing, or unfolding; a plan or protocol for carrying out or accomplishing something; the arrangement of elements or details in a product or work of art." Today, we design large, complex process, systems, and services, and we design organizations and structures to produce them. Design has changed considerably since our remote ancestors made the first stone tools.

At a highly abstract level, Herbert Simon's definition covers nearly all imaginable instances of design. To design, Simon writes, is to "[devise] courses of action aimed at changing existing situations into preferred ones" (Simon, *The Sciences of the Artificial*, 2nd ed., MIT Press, 1982, p. 129). Design, properly defined, is the entire process across the full range of domains required for any given outcome.

But the design process is always more than a general, abstract way of working. Design takes concrete form in the work of the service professions that meet human needs, a broad range of making and planning disciplines. These include industrial design, graphic design, textile design, furniture design, information design, process design, product design, interaction design, transportation design, educational design, systems design, urban design, design leadership, and design management, as well as architecture, engineering, information technology, and computer science.

These fields focus on different subjects and objects. They have distinct traditions, methods, and vocabularies, used and put into practice by distinct and often dissimilar professional groups. Although the traditions dividing these groups are distinct, common boundaries sometimes form a border. Where this happens, they serve as meeting points where common concerns build bridges. Today, ten challenges uniting the design professions form such a set of common concerns.

Three performance challenges, four substantive challenges, and three contextual challenges bind the design disciplines and professions together as a common field. The performance challenges arise because all design professions

1. act on the physical world,
2. address human needs, and
3. generate the built environment.

In the past, these common attributes were not sufficient to transcend the boundaries of tradition. Today, objective changes in the larger world give rise to four substantive challenges that are driving convergence in design practice and research. These substantive challenges are

1. increasingly ambiguous boundaries between artifacts, structure, and process;
2. increasingly large-scale social, economic, and industrial frames;
3. an increasingly complex environment of needs, requirements, and constraints; and
4. information content that often exceeds the value of physical substance.

These challenges require new frameworks of theory and research to address contemporary problem areas while solving specific cases and problems. In professional design practice, we often find that solving design problems requires interdisciplinary teams with a transdisciplinary focus. Fifty years ago, a sole practitioner and an assistant or two might have solved most design problems. Today, we need groups of people with skills across several disciplines and the additional skills that enable professionals to work with, listen to, and learn from each other as they solve problems.

Three contextual challenges define the nature of many design problems today. While many design problems function at a simpler level, these issues affect many of the major design problems that challenge us, and these challenges also affect simple design problems linked to complex social, mechanical, or technical systems. These issues are

1. a complex environment in which many projects or products cross the boundaries of several organizations, stakeholder, producer, and user groups;
2. projects or products that must meet the expectations of many organizations, stakeholders, producers, and users; and
3. demands at every level of production, distribution, reception, and control.

These ten challenges require a qualitatively different approach to professional design practice than was the case in earlier times. Past environments were simpler. They made simpler demands. Individual experience and personal development were sufficient for depth and substance in professional practice. While experience and development are still necessary, they are no longer sufficient. Most of today's design challenges require analytic and synthetic planning skills that cannot be developed through practice alone.

Professional design practice today involves advanced knowledge. This knowledge is not solely a higher level of professional practice. It is also a qualitatively different form of professional practice that emerges in response to the demands of the information society and the knowledge economy to which it gives rise.

In his essay "Why Design Education Must Change" (from *Core77*, November 26, 2010), Donald Norman challenges the premises and practices of the design profession. In the past, designers operated on the belief that talent and a willingness to jump into problems with both feet gives them an edge in solving problems. Norman writes:

> In the early days of industrial design, the work was primarily focused upon physical products. Today, however, designers work on organizational structure and social problems, on interaction, service, and experience design. Many problems involve complex social and political issues. As a result, designers have become applied behavioral scientists, but they are woefully undereducated for the task. Designers often fail to understand the complexity of the issues and the depth of knowledge already known. They claim that fresh eyes can produce novel solutions, but then they wonder why these solutions are seldom implemented, or if implemented, why they fail. Fresh eyes can indeed produce insightful results, but the eyes must also be educated and knowledgeable. Designers often lack the requisite understanding. Design schools do not train students about these complex issues, about the interlocking complexities of human and social behavior, about the behavioral sciences, technology, and business. There is little or no training in science, the scientific method, and experimental design.

This is not industrial design in the sense of designing products, but industry-related design, design as thought and action for solving problems and imagining new futures. This MIT Press series of books emphasizes strategic design to create value through innovative products and services, and it emphasizes design as service through rigorous creativity, critical inquiry, and an ethics of respectful design. This rests on a sense of understanding, empathy, and appreciation for people, for nature, and for the world we shape through design. Our goal as editors is to develop a series of vital conversations that help designers and researchers to serve business, industry, and the public sector for positive social and economic outcomes.

We will present books that bring a new sense of inquiry to the design, helping to shape a more reflective and stable design discipline able to support a stronger profession grounded in empirical research, generative concepts, and the solid theory that gives rise to what W. Edwards Deming described as profound knowledge (Deming, *The New Economics for Industry, Government, Education*, MIT, Center for Advanced Engineering Study, 1993). For Deming, a physicist, engineer, and designer, profound knowledge comprised systems thinking and the understanding of processes embedded in systems, an understanding of variation and the tools we need to understand variation, a theory of knowledge, and a foundation in human psychology. This is the beginning of "deep design"—the union of deep practice with robust intellectual inquiry.

A series on design thinking and theory faces the same challenges that we face as a profession. On one level, design is a general human process that we use to understand and shape our world. Nevertheless, we cannot address this process or the world in its general, abstract form. Rather, we meet the challenges of design in specific challenges, addressing problems or ideas in a situated context. The challenges we face as designers today are as diverse as the problems that clients bring us. We are involved in design for economic anchors, economic continuity, and economic growth. We design for urban needs and rural needs, for social development and creative communities. We are involved with environmental sustainability and economic policy, agriculture competitive crafts for export, competitive products and brands for microenterprises, developing new products for bottom-of-pyramid markets and redeveloping old products for mature or wealthy markets. Within the framework of design, we are also challenged to design for extreme situations; for biotech, nanotech, and new materials; for social business; as well as for conceptual challenges for worlds that do not yet exist (such as the world beyond the Kurzweil singularity) and for new visions of the world that does exist.

The Design Thinking, Design Theory series from the MIT Press will explore these issues and more—meeting them, examining them, and helping designers to address them.

Join us in this journey.

Ken Friedman
Erik Stolterman

Editors, Design Thinking, Design Theory Series

Preface and Acknowledgments

This book articulates a vision of interpretive design research by analyzing the history of an empathic design research group in Helsinki, Finland. The book argues that interpretive design research is a versatile approach to design that can offer answers to the most important variables design needs to cover: human beings, design itself, techniques, and technology.

How to define interpretive design research? Like any practice, it escapes definition, but it has a few characteristics other approaches to design research do not share. It builds explicitly on an interpretive way to study human beings for research. Its sine qua non is a theory of meaning that drives human action. Instead of being motivated by drives, cognition, emotions, or social forces like identities, this theory claims that to understand humans, we need to understand how they give meanings to things—themselves, people around them, things around them, or abstractions like "Paris" or "design." These words mean different things to different people, and they are ultimately defined in the context in which they occur. Interpretive theory explicates how meanings are created, maintained, and transformed in human interaction and claims that human action is dependent on these meanings. A working definition of interpretive design research is that it is research that builds consciously on an interpretive theory of human action but translates it into concepts, research techniques, design frameworks, and sometimes design processes and artifacts.

The origins of this perspective are from the late 1990s. It was an era that changed design research rapidly. During a period from 1995 to 2005, design researchers in several universities and companies started to build research on design practice and shifted research from archives into the studio. This change happened in several places simultaneously. Especially noteworthy were communities in Chicago, Silicon Valley, the Netherlands, England, Italy, and the Nordic region of Europe. As this book shows, the interpretive approach of empathic design articulated a unique vision of what design research is/could be in the context of alternatives, including ethnomethodology in Xerox PARC, symbolic anthropology at Intel, ecological psychology in the Netherlands, situationism and critical theory in London, and participatory design in Sweden and Denmark. By now, the focal point of design research has shifted from this base toward a systemic perspective, but the base is still influential and important, and the message of focusing on designing for human beings is as important as ever.

I will focus on one research community and one research program in this community. This is an unusual approach in design literature and warrants a few lines about the method I am using.

To my knowledge, this is the first book about the development of design research that makes a sustained argument by analyzing one user-centered design research program. The method is different from the usual method authors like Redström (2017), Tharp and Tharp (2018), Wakkary (2021) or I (Koskinen et al. 2011) have used to articulate frameworks for theory construction, discoursive design, and designing-with nonhumans. These authors have canvassed many studies from design literature to support their argumentation *a forteriori*. The frameworks by these authors come from a space they have put together to support their argumentation. At the outset, it looks like the frameworks are general because they canvass the whole world. Yet, it can also be argued that their power is limited by the fact that they skip over contextual specifics, be they personalities, academic practices, funding models, or the local composition of industry.

In comparison, when a framework is built from a consistent case, we face constraints these authors could assume away. The same forces have created questions and shaped several pieces of research: for instance, the emergence of user experience around the turn of the century, the shift to social media after 2006, or a change in the research strategy of the faculty. By analyzing how researchers have turned these constraints into opportunities, we can trace patterns in the program and use these patterns to describe how the framework behind the program has shaped its story. In this book, I shall look at one of these patterns, the interpretive framework, and study its behavior in detail to see how it has shaped individual studies, and how it has shaped subsequent studies in new circumstances. In the process, we learn about the limitations of the framework: as the book shows, many things in the program were relevant for a while but became dated a few years later.

This is probably the main benefit of the method of this book: it tracks the behavior of the framework while being sensitive to its context. The method makes the analysis presented in this book richer than the method of those *a forteriori* studies—my previous work included—that, *ceteris paribus*, could skip over the contextual aspects of research.

The method has many precedents if we cross the line from design research to design. For example, some of my favorite works include Markus Rathgeb's (2006) study of Otl Aicher in the context of German industry, the School of Ulm, and the Olympics in München; and Issey Miyake's exhibition *The Work of Miyake Issey* in the National Art Center, Tokyo, which traced his work as a series of innovations from sculptural forms and one-cloth pieces to pleats (Issey 2016). There are also fantastic exhibition catalogues. Some of my favorites are Andrea Branzi's research into foundational metaphors of Italian design in *Che cosa è il design italiano? Le sette ossessioni* (2008) and Silvana Annicchiarico and Beppe Finessi's examination of reactions to austerity in Italian design in *Il design italiano oltre le crisi* (2014), both published in a series of annual exhibitions in Triennale di Milano. Other studies focus on companies (Heskett 1989; Kicherer 1990) and schools (Spitz 2002; Schouwenberg and Staal 2008; McIntyre 1995). A recent paper that has traced the development of the design community in Delft touches on research in parts in the context of a university department (Voûte et al. 2022).

Some research books give even better precedents. For example, Pelle Ehn's (1988) doctoral thesis has an overview of the history of participatory design in a story that tried to find an explanation as to why rough mock-ups worked better as design tools than scientific diagrams. Brian Dixon's *Dewey and Design* (2020) attempted to expand the design community's knowledge of Dewey but did not trace Dewey's impact in design in one community. In *Design Things*, Thomas Binder and his colleagues studied a group of design students in Vienna working on an EU-funded project to understand how designers use sketches and mock-ups. They also tried to build a Latourian and etymological framework out of the case (Binder et al. 2011). Their book contributed to understanding physical practices in design and it is still perhaps the closest methodic analogy to the current book, but it focused on students whose job was to design concepts rather than create knowledge.

Another recent book that builds on a curated collection of cases is *Drifting by Intention*, a book I wrote with Peter Krogh (Krogh and Koskinen 2020), which analyzed epistemological concepts in a body of about sixty doctoral theses mostly from two main research programs, participatory design in Denmark and user-centered design in Helsinki. The current book takes this method a step further by focusing on one community, isolating one way of thinking in this community, and studying its contour over the years.

The main difference to these books is that the hero of the current book is not an individual, as in Dixon's book, or a community, as in Binder's book or in the book I authored with Krogh. The hero is an interpretive framework that has its origins in Herbert Blumer's (1969) formulations. It is the framework different researchers have applied to their problems and reinterpreted when they found it wanting.

Does the method of the book limit its applicability in other contexts? You can argue that the method makes it applicable only in towns like Helsinki, which has a sophisticated system of government, high-tech industry, and research funding systems, as well as one of the world's leading design schools. Is Helsinki a smaller version of Milan that has a unique "design system" (Bertola et al. 2009), an ecology of designers, shops, galleries, schools, publishers, and museums that makes it hard if not impossible to replicate elsewhere? Several things speak against this argument. The ecosystem in Helsinki has similarities to several countries in Europe, Canada, and Australia; research programs have cognitive autonomy; and the program's focus on ordinary action is robust. We pass the salt when someone asks for it at the dinner table, just as our grandparents did and our grandchildren will: there are constants in human life and theories and methods to capture them (Koskinen, Kurvinen, and Lehtonen 2002, 9).

Perhaps more important, the method has precedents in many well-established disciplines outside design. For example, political scientists routinely compare electoral systems between countries—d'Hondt in one country, first-past-the-post in another, Sainte-Laguë in a third—even though countries are very different at heart. The trick is the unit of analysis—focusing on the performance of the system, its function, rather than on details. In this book, the focus is on one framework. The difference to fields like electoral studies, of course, is that there are no detailed studies yet about, say, how ecological psychology has provided a consistent thread to research in Eindhoven, pragmatism in Carnegie-Mellon University, or systems theory in Milan (but see Overbeeke 2007; Koskinen, Forlizzi, and Battarbee forthcoming; Manzini 2015). It would be impossible to study electoral systems without cases; design research is no different. Cases like the one I am dealing with in the present book will hopefully be the foundation of comparisons that may one day correct my interpretations and also provide a better way to understand things that shape design research.

This book has been made possible by an extraordinary group of design researchers. I must single out five of them: Turkka Keinonen and Simo Säde as trailblazers and Tuuli Mattelmäki, Katja Battarbee, and Esko Kurvinen as key researchers. In addition, there are several designers without whom the group would have not survived. Juhani Salovaara helped to create the group, and he also taught his practical methods to it; Raimo Nikkanen supported design research at the University of Art and Design Helsinki (UIAH, now Aalto University); and Eero Miettinen's speech in UIAH's Lume center in 2001 sparked the idea of seeing design as a form of interpretive research.

I need to thank Petra Ahde-Deal, Heidi Paavilainen, Jung-Joo Lee, Andrea and Marcelo Júdice, Kirsikka Vaajakallio, Katja Soini, Karthikeya Acharya, Meri Laine, Juha Järvinen, Yiying Wu, Jussi Mikkonen, Maarit Mäkelä, Krista Kosonen, Pirjo Kääriäinen, Salu Ylirisku, Susanna Jacobson, and Kirsi Niinimäki. My apologies for those colleagues whom I have not mentioned. Colleagues who deserve thanks are Anne Burdick, Lisa Nugent, and Sean Dohanue, who hosted first me and later Tuuli Mattelmäki in 2006–2007 at ArtCenter College of Design in Pasadena, California. Kari Kuutti, Martti Mäntylä, and Giulio Jacucci taught the group the importance of user research in computer science. Mika Pantzar taught the group to see its work in the context of science studies. Jack Whalen has been a constant source of ethnomethodological insights and stories, and a few emails with Ron Wakkary gave me a way to finish this book. Janet McDonnell and Stephen Scrivener opened the history of codesign in England to me, and Francesca Rizzo in Milan. Thomas Binder, Johan Redström, and Peter Krogh connected the group to an emerging Nordic research community. Thanks to Pieter Jan Stappers, John Cain, Caroline Hummels, Anna Meroni, and Kristina Höök for stories of their research programs. The reviewers of this book were knowledgeable and strict and gave excellent suggestions. John D'Souza revised the language of the book, and Christine Tsin created its graphics.

I must mention the pioneering work of Jane Fulton Suri and Liz Sanders as inspirations. Another key inspiration is Jodi Forlizzi. She provided the group with a crucial cue that has shaped its path since 2000. I hope these three will enjoy this book.

This book is a story of the ramifications of an interpretive foundation, not a history of the whole research program. Readers ought to keep in mind that I have been a part of the story. I have checked the stories of others in as many ways and as best as I could, but they are inevitably my versions. I have tried to recover emails from old servers and accounts, but much of the history has been lost in the world of electronic communications and long-gone servers. The paper trail is also largely missing because research notes and old drafts are seldom archived. My main unpublished sources are old manuscripts and paper drafts written by people mentioned in this book, and my own notes and eyewitness memories.

June 20, 2022,
Ilpo Koskinen

Smart Products

Usability of Smart Products

Maypole

One-Dimensional
Usability

Cardboard Mock-ups

The Future of
Digital Imaging

eDesign

Empathic Design

Design Games

IKE

Mobile Image

Design Studio
in the Field

Luotain

Radgoing

Co-experience

Collaborative
Design

Design Probes

Prototyping
Social Action

Women
and Jewelry

You are Important

Design for Hope

Dwelling with Design

Pasadena

Morphome

Electrome

IP08

Art of
Research

Prototyping
Interactions

SPILE Project

Design Research
through Practice

Sites and Plants

Against Method

Memories in Clay

Drifting by Intention

Folk Tradition

Runkohuituja

What Happened

Novapro

From Disposable
to Sustainable

DWoC

Lost in Woods

1 Interpretive Turn in Design Research

The rise of information technology created a need for user-centered design approaches in the 1990s. Some of these approaches were interpretive. They saw human beings as creatures who act on meanings.

One of the interpretive approaches came from the former University of Art and Design Helsinki (UIAH), now Aalto University, where Juhani Salovaara, Turkka Keinonen, and Simo Säde developed methods for studying usability.

When Tuuli Mattelmäki, Katja Battarbee, Esko Kurvinen, and I joined the group in 1999–2000, the group moved to user experience. The group built an interpretive, empathic approach to it on three principles outlined by the sociologist Herbert Blumer: people act on meanings that are created and modified in interaction with others.

The approach was developed in several projects: *Maypole* (1997–1999), *eDesign—Design for Emotional Experience in Product Use* (1999–2001), and *The Future of Digital Imaging* (1999–2000). Over the next twenty years, the group grew and its research expanded to several new areas.

This chapter introduces the group and puts it in international and theoretical context.

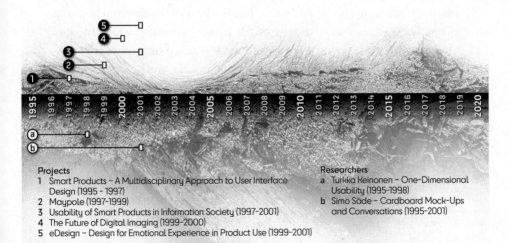

Projects
1 Smart Products – A Multidisciplinary Approach to User Interface Design (1995 - 1997)
2 Maypole (1997-1999)
3 Usability of Smart Products in Information Society (1997-2001)
4 The Future of Digital Imaging (1999-2000)
5 eDesign – Design for Emotional Experience in Product Use (1999-2001)

Researchers
a Turkka Keinonen – One-Dimensional Usability (1995-1998)
b Simo Säde – Cardboard Mock-Ups and Conversations (1995-2001)

Design had a lively research world in the beginning of the 1990s. Design researchers were familiar with ergonomics, information theory and semantics, engineering processes and methods, and management theories. The field of design history was ready to leave its home in art history. A larger upheaval was on its way, however. Over the last three decades, design research has gone through a user-centered turn. It has sought ways to improve the knowledge of human beings and their relationship to technology. Today, design research includes thousands of researchers working on visual design, products, spaces, interactive designs, services, social topics, policy, and, increasingly, nonhuman things of many kinds. A good deal of this change owes a debt to the user-centered turn.

Behind this turn have been several drivers. Perhaps most important, design research has been driven by the rise of information technology into the center of modern society. It has created new demands on designers since the 1980s, which witnessed the rise of user-centered design. When the old world of mainframes gave way to personal computing, understanding users became a necessity. The former was operated by trained personnel, and the latter by users at home. By the early 1990s, the internet was morphing from a scientific tool into the World Wide Web, mobile phones started to enter our pockets and bags, and microchips gave products intelligence. Interfaces for these devices could not be built using the traditional design tools.

Since then, we have seen the rise of Web 2.0, and now Web 3.0, the birth of modern search engines and business models based on clicks, social media and microblogs, ubiquitous computing, and wearables, to mention just a few. The design offices have seen a thorough shift from analog to digital methods. We have also seen hiccups, like the bust of the dot-com boom in the early 2000s, as well as the rise of the sharing economy, which has created products and services to ease our lives but also broken industrial relations, created new ways to exploit infantile fantasies, and deserted some of the world's greatest city centers.

Design research has also been shaped by issues even larger than information technology, including the rise of identity politics from a fringe phenomenon into the very center of society, the shift of global warming from a scientific question into an existential planetary-level threat, and the rise of inequality associated with a confluence of new technology, the globalization of economy, shrinking government and unions, and new business models that have raised ethical challenges by shifting the balance of economy from the have-nots to the haves. The world in which design researchers operate is not what it was in 2000, a time that feels innocent now.

For design researchers, the last three decades have been turbulent. Design was losing its old foundations in this context. The tools from the postwar industrial era were designed to work on physical things, and although the new digital tools of the trade helped the design occupations upgrade their skills, new methods were needed when the challenge was to envision new things and human practices rather than perfect forms and manufacturing techniques. Design researchers started to develop a variety of user-centered techniques from usability to design ethnography, and user experience became a common concern for practitioners and academic researchers. For example, design research has progressed theoretically. Several researchers have introduced psychological, philosophical, critical, artistic, sociological, anthropological, systemic, and philosophical perspectives to design (e.g., Overbeeke 2007; Sanders 2000; Forlizzi and Ford 2000; Dunne and Raby 2001; Dunne and Gaver 2001; Crabtree 2003; Manzini 2015; Dixon 2020). The field has also seen methodic progress. Every designer today learns about contextual inquiry, cultural probes, experience prototypes, and many other generative methods that usually had their roots in design practice but were turned into research tools in several leading design communities. These methods have been complemented by fictional methods and methodological studies (e.g., Beyer and Holtzblatt 1998; Gaver, Dunne, and Pacenti 1999; Hummels 2000; Buchenau and Fulton Suri 2000; Sanders 2000; Bleecker 2009; Koskinen et al. 2011). By now, design researchers have developed dozens of frameworks for specific design problems (e.g., Battarbee 2004; Frens 2006; Ludvigsen 2006).

Looking back, the progress of the field has been remarkable. A good deal of it can be attributed to long-running research programs. For example, for almost three decades, one thread of research at the Technical University of Eindhoven has been focused on turning bodily perception that does not require thought into interfaces (Overbeeke 2007). This program has generated models for virtual reality (Djajadiningrat 1998), imaginative forms for alarms (Wensveen 2004), novel interfaces for cameras and lights (Frens 2006; Ross 2008), as well as data (van Kollenburg and Bogers 2018). At the core of these studies has been J. J. Gibson's ecological psychology, which has inspired the topics and approach of individual researchers over the years, but the topics sometimes have come from personal interests, sometimes from industry.

Theories like Gibson's have been central to many leaps of the field, but the way in which they have shaped the field has not been described systematically. This is the gap that this current book addresses: it tells the story of a research program that has built design research around a few principles. At the core of this book is the idea that people create meanings with other people, and it is these meanings that drive the uses of design and technology. Design researchers are familiar with an interpretive approach from ethnomethodology (Suchman 1987), phenomenology (Dourish 2002; Pallasmaa 2009), symbolic anthropology (Salvador, Bell, and Anderson 1999) and activity theory (Nardi 1996). Most of this literature comes from computer science, however, and it does not address the key questions of this current book. What would interpretive design research look like? What kinds of elements does it have, how can it contribute to design research, and can it contribute beyond its user-centered basis? These questions are answered through a case study of empathic design, but we need to put it in context first.

From Usability to User Experience

The story of this book begins in the 1990s. Information technology was becoming ubiquitous at that time. It created a lot of work for designers, but their traditional tools were ill equipped for dealing with it. An industrial tool was around the corner, however. The notion of usability gave them tools to design easier-to-operate user interfaces for the web, mobile phones, and other smart, software-operated devices like medical instruments, heart rate monitors, and wrist-top computers for sailing, skiing, surfing, and mountaineering.

Usability had grown into a research field by the second half of the 1980s, and its culmination was Jacob Nielsen's *Usability Engineering* (1993). In an editorial to the *Journal of Usability Studies* in 2007, Joe Dumas chronicled the history of the usability profession and traced it to dissatisfaction with attempts to apply methods from the behavioral sciences to design interfaces. The early 1980s saw the emergence of human-computer interaction (HCI) as a research field that borrowed its methods mostly from psychology. There were few academic programs in HCI, and those programs taught methods and theories from cognitive psychology and applied ergonomics (Sanders and Dandavate 1999, 87). They had several benefits. When researchers viewed humans in terms of cognitive processes like perception, memory, judgment, and decision-making, they could draw testable propositions that reduced the need for endless testing. For example, they could predict that if it takes a long time before novices can understand a user interface, and they cannot recognize their navigation errors; they will dislike these kinds of interfaces.

However, these methods faced problems on the practical side, where the main approach was to build design guidelines, many of which were becoming so complicated that they were losing their practical value. In contrast, the emerging usability profession had a practical, nuts-and-bolts approach to design. Instead of guidelines, usability engineering was built on heuristics and walkthroughs (Nielsen and Molich 1990; Polson and Lewis 1990). Dumas dates the birth of usability in the early 1980s to the work of John Bennett at IBM. As he notes, the ethos that drove early usability was different from HCI. Its foundation was early goal setting, prototyping, and iterative evaluation rather than rigorous scientific experimentation. In addition, it moved research from the lab to a work context and favored the integration of usability teams into product design teams. In many ways, usability engineering liberated usability from guidelines with methods that worked better under the tight timelines of industry and targeted practical problems (Dumas 2007, 55).

Emotions and Pleasure as Design Topics

By around 2000, it was becoming clear that usability research was facing some natural limits. It helped designers to eliminate problems in user interfaces, but it said little about what to design or how to design it. It was research after the fact—to be able to study a peripheral like a mouse, you must have it first—and did little to help the creative aspect of design. A way around this problem was emerging from business literature, however. Joseph Pine and James Gilmore's *The Experience Economy* (1999) was a particularly important resource. Their book painted a broad-strokes picture of an economy in which most value was produced in experience industries that catered to customers' feelings. The experience economy stood in stark contrast to earlier forms of economy in which value was created in the primary sector, industry, and services. Another reference was the Danish futurologist Rolf Jensen, whose book *The Dream Society* (1999) claimed that imagination and storytelling, rather than information, were the future.

A few writers had brought these discussions into design as well. In *Design Management Journal*, Darrell Rhea advocated as early as 1992 that designers need to focus on experience, and when Dorothy Leonard and Jeffrey F. Rayport argued in the *Harvard Business Review* in 1997 that empathic design is important in sparking innovation, some designers felt at home in business literature for once. The key words in this literature were "emotions" and "experience," and they reintroduced Alvin Toffler's futurology, Jean Baudrillard's postmodernism, and Gerhard Schultze's sociology into design (Rhea 1992; Leonard and Rayport 1997; Pine and Gilmore 1999; Baudrillard 1976; Toffler 1980; Schultze 1992).

The message of the experience literature became clear for design practitioners over the next few years, but the literature was theoretical and left designers almost empty-handed when it came to how to change design. There was an emerging body of design literature that specifically tackled this problem, however, and it built on the most advanced practices of the era.

In the 1990s, a handful of design firms had begun to switch their focus from usability to emotions and, later, experience. In this process, they began to use the word "empathy." At IDEO, Jane Fulton Suri, Leon Segal, and Marion Buchenau built methods for studying emotions and described their approach as empathic. At IDEO's London office, Alison Black also spoke about the need for an empathic approach. Another source was Liz Sanders from Columbus, Ohio. These designers shifted the discussion in two important ways. First, their central concept was empathy, not need. The very word "empathy" clearly implies that designers must see the world through the eyes of the people for whom they are designing. Second, their research proved that it is possible to do research about emotions in industry with qualitative, design-based methods that were appropriated from design practices rather than the sciences or the social sciences. (See Fulton 1993; Dandavate, Sanders, and Stuart 1996; Segal and Fulton Suri 1997; Black 1998; Sanders and Dandavate 1999; Buchenau and Fulton Suri 2000.)

By the end of the 1990s, emotions had become a hot topic in design research. For example, at the end of the decade, the Affective Computing Research Group at the Media Lab at the Massachusetts Institute of Technology (MIT) pioneered a sophisticated approach to emotions under the term "affective computing." Rosalind Picard's book *Affective Computing* (1997) became a reference point in engineering. She claimed that computer scientists and engineers need to find ways to recognize and understand emotions in order to be able to give emotions and emotional expressions to computers. For example, human behavior could be observed through computer vision

and sensors. By collecting and processing information from sensors, computers could be taught not only to know what people tell them, but also to sense
emotions that could be used as an input in developing products and systems.

These interpretations found a receptive audience in the emerging mobile
industries in Europe. Nokia in Helsinki and Ericsson in Stockholm were at the
forefront of context-aware computing. Phones could be thought of as a vast
sensory network that could be used for collecting information about where
people are, noise levels around the phone, and other issues. What if phones
could also track users' emotions with sensors measuring heartbeat and galvanic skin response? What if this information could be used in designing
phones and services? The argument was also the cornerstone in the trans-
formation of Alessi from a maker of stainless steel cooking equipment to a
playful, emotionally grounded manufacturer of products that brought joy to
the kitchen.

For high-tech companies, this vision was attractive and powerful, but it
downplayed the emotional side of human beings. Patrick Jordan from Philips Design examined emotion through hedonic psychology (see also Blythe
et al. 2002). He studied a group of people in Glasgow, Scotland, to learn about
the limits of usability. His paper "Human Factors for Pleasure in Product Use"
(1998) argued that displeasure with products and pleasure with products are
separate factors and need distinctive design approaches. Design had to go
beyond usability to create products that are a pleasure to use, not merely
efficient:

> Usability is a central factor in whether or not a product is pleasurable to use. How-
> ever, the issue of pleasure in product use also goes significantly beyond usability.
> The emotions felt when using pleasurable/displeasurable products are potentially
> more wide-ranging than just satisfaction/dissatisfaction, and the properties of a
> product which influence how pleasurable/displeasurable it will be to use do not
> only include the property of usability. To fully represent the user in the product
> creation process, the human factors specialist should look both at and beyond
> usability to create products that are a positive pleasure to use. (Jordan 1998, 32–33)

Jordan's message was attractive to design researchers in the Netherlands
and Scandinavia, but his theoretical beliefs were not always shared in these
communities. For example, by linking pleasure to needs, he joined a line
of psychologists going all the way back to Sigmund Freud's psychoanalysis.
While some researchers followed psychoanalysis, most were reading Abraham Maslow (1946) instead. Maslow classified human needs into a loose

hierarchy that begins with biological needs and progresses to higher needs like aesthetics and justice. Others built on cognitive psychology (Hassenzahl 2004) and Gibson's ecological psychology (Overbeeke 2007).

Although Maslow's psychology was simple enough, it had its problems. It made more sense at the lower levels of needs than at the higher end. If I am sleep deprived, I need to sleep, and if I am dehydrated, I need to drink. If I am yearning for an aesthetic experience to enrich my life, it is not at all clear what I need to do. Do I "need" to listen to György Ligeti or John Adams, buy a chair designed by Ray and Charles Eames, or buy a ticket to an *Ikebana* show? What if I am a designer whose job is to turn visiting a high-end restaurant into a memorable experience or to give local flavor to a new subway station? Would it be better to use another word besides "need"?

User Experience and Pragmatism

A term that avoided these problems was around the corner. It was the notion of user experience, which had become popular in the digital industries that the World Wide Web had spawned in the second half of the 1990s (Shedroff 1991). A good deal of the literature on user experience was practical. For instance, in 1998, John Cain described how the E-Lab approached experience by breaking it into sociocultural systems, patterns and routines of action, and things that people use. Rick Robinson built the model, and it boiled down to three components. "Think" stood for ideas, beliefs, attitudes, and expectations, but also the sociocultural systems that inform them. "Do" captured patterns and routines of action. Finally, "use" captured things that people use and the impact that these things have on ideas, beliefs, attitudes, and expectations (Cain 1998, 12).

Working at Carnegie Mellon University, Jodi Forlizzi and Shannon Ford's paper "The Building Blocks of Experience: An Early Framework for Interaction Designers" (2000) introduced a pragmatist framework of user experience. Their framework made no distinction between emotions and cognition, but it started from the stream of experience following John Dewey's (1980) pragmatism, which they had learned in Richard Buchanan's seminars in Pittsburgh. In this framework, humans continuously experience things. Some of these experiences form the stream of mere experience: sounds, smells, thoughts, and words that we hear in our life subconsciously without paying much attention to them. In contrast to mere experience, "an experience" is lifted out of this stream. It becomes noticeable, memorable, and people may tell stories about it. It transcends those barely noticeable things that we come in touch with through our senses, and it is something we think and talk about. As Dewey wrote, when you say that a dinner in Paris was "an experience," you know that it was more than just a regular meal. You know that you will remember it for years to come and tell stories about it. Most other European languages distinguish these qualities with two words, including *Erfahrung* and *Erlebnis* in German.

With their paper, Forlizzi and Ford gave a philosophical grounding to user experience. By shifting user experience to pragmatism, they put design into a much broader and richer intellectual environment than Nielsen's usability research or Jordan's hedonic psychology. Experiences must be studied in the larger context of life rather than in isolation in a laboratory. Having a sound philosophical basis in Dewey opened many research avenues, as Dalsgaard (2009) and Dixon (2020) amply show.

Forlizzi and Ford's paper was fully in line with Dewey's philosophy and opened plenty of questions for design researchers. When seen in pragmatic terms, there was no need to split the human world into knowledge and emotion. The shift to pragmatism broadened research from user interfaces to life in general. Why stick to emotions and knowledge rather than interactions and those social practices in which interactions take place? Why should we not situate these practices in institutions? This was consistent with how practicing designers understood experience. Forlizzi and Ford (2000) proved that it is possible to give an academic interpretation of industrial terms and thus get access to intellectual resources well beyond practice. Importantly, she was an illustrator by training. Her model was abstract, but perhaps for this reason, it felt right for design readers. Her paper "Understanding Experience in Interactive Systems" (Forlizzi and Battarbee 2005) is perhaps the most highly cited user experience paper written by designers.

Forlizzi's framework was not the only pragmatic interpretation of user experience in design research, but it had its benefits. For example, in the book *Technology as Experience*, McCarthy and Wright (2004) described user experience in terms of compositional, spatiotemporal, emotional, and sensual "threads." They also organized these into a process-based model that started from connecting and went to interpreting, reflecting, appropriating, recounting, and finally anticipating. This framework made distinctions that were hard to reconcile with the pragmatist principle that told philosophers to study experience without preconceived concepts and distinctions. Forlizzi and Ford's framework, in contrast, avoided commitment to preconceived ideas and categories and resonated well with the idea that design ought to remain open to possibilities rather than build on theoretical frameworks.

Usability Morphs into Empathic Design

The basis of this book is empathic design as it has been practiced at Aalto University in Helsinki, Finland. It is a research program that has generated more than thirty monographs and several hundred articles and conference papers since its start around 1997–1998. What makes its story worth telling is that it is one of the few design research programs that has been consistently developing and progressing into new topics and technologies. It has also been well received in industry and universities. All the while, its foundation has become interpretive. It is this foundation that has made it unique in the already-crowded field of design research.

The key early members of the group were Tuuli Mattelmäki, Katja Battarbee, Esko Kurvinen, and myself. The ground was laid by Turkka Keinonen, now a professor at Aalto University. His doctoral thesis *One-Dimensional Usability* (1998) studied the relationship of usability and consumer preferences. He saw that usability is an important determinant of preference and varies with knowledge of user interfaces. He moved to industry after his thesis. Another early trailblazer was Simo Säde, an industrial designer who was working on usability methods with software engineers. After publishing his doctoral thesis *Cardboard Mock-Ups and Conversations* (2001), he also moved to industry. Keinonen and Säde studied the usability of smart products—small, handheld products with a significant software component, including things like heart rate monitors and mobile phones—and organized their work under the Smart Products Research Group (SPRG) with Professor Juhani Salovaara.

After Keinonen left the University of Art and Design (now Aalto University) in 1998, it hired Mattelmäki and Battarbee, two doctoral students with a background in industrial design. These two worked together on the European Union project *Maypole* (1997–1999), led by the Netherland Design Institute. It developed, tested, refined, prototyped, and carried out field studies of mobile multimedia devices in local communities (Hofmeester and Gijsbers 1999; Hofmeester and de Charon de Saint Germain 1999). The most important industrial partners of the project were Nokia Research Center and IDEO Europe. The project took these two researchers to the first Design and Emotion conference in Delft, the Netherlands, in 1999. In the Netherlands, they got in touch with the Dutch design research community. They also met Liz Sanders of Fitch (and later SonicRim), who was also teaching at Ohio State, and Bill Gaver, a former Xerox researcher who was working at the Royal College of Art in London. In 1999, the group also hired Kurvinen, Turo-Kimmo Lehtonen, and me to study the future of digital imaging.

In 2000, the group's interest still lay in designing user interfaces and products that build on software. As the group grew, however, its work became more diverse. It gave birth to a larger research program in industrial design and later in several other areas of design. The group built on several theories and philosophical schools, but the *eDesign* project formulated its interpretive foundation, which gave the program a distinctive approach that has carried it for over twenty years.

The group had started to face new types of requests from industry, however. Industry was asking them to work on lifestyles, feelings, hobbies, and minuscule behaviors like how things are carried. Traditional design frameworks and methods that had been built to solve problems in designing physical products were not able to address these questions: they were not problems in the traditional sense of the term. It was also clear that these questions were not going to be a fad that would pass quickly; they stemmed from a profound technological change driven by Silicon Valley and, in Scandinavia, at Nokia in Finland and Ericsson in Sweden. New methods were needed to design these technologies, and although usability was useful for weeding out things that ruined user experience, there was also a need to go beyond usability.

The shift had a familiar undercurrent. Design had gone through a massive digitalization in the early 1990s. Computer-aided design (CAD) and computer-aided manufacturing (CAM) technologies were becoming commonplace, and they were rapidly replacing traditional tools. Yet the hard-won tools that had reshaped several design industries were ill equipped to answer questions emerging from these new digital industries. How do you generate a CAD model of how people carry things? How can you model brawling? How do you turn health scares about electromagnetic radiation into products? How do you model friendship in the city? These kinds of questions were discussed in design studios, and old hands in the industry developed their workarounds. Sketches, technical drawings, mock-ups, scale models, site plans, and advanced joinery were useful, but they were not the best tools for tackling these questions.

Questions like these put people into the center of the design process. For decades, industrial designers had thought of themselves as spokespersons for people in technology, but few knew how to study humans at the level that the new industries needed. Over the decades, the best human-centered methods had come from ergonomics (Tilley and Dreyfuss 2002; Salovaara 1985; Ball 2011) and from product semantics (Butter 1989; Vihma 1995, 2010; Karjalainen 2004; Krippendorff 2006), but both were inadequate for responding to the emerging challenges. As design was shifting from a solution-oriented to a question-oriented discipline, its new problem was what to design rather than how: how to capture something from the human world in product development rather than perfecting forms from the industrial era.

The Interpretive Foundations of Empathy

User experience had become the main focus of the members of the group by 2000. The group liked the term because it had philosophical depth and was used in information technology industries. Theoretically, they turned toward pragmatism while their practical approach built on design practice. They had learned the practical approach firsthand from Turkka Keinonen, Alison Black, and Jane Fulton Sure in the *Maypole* and *eDesign* projects described earlier. Professor Juhani Salovaara was another role model: in his studio, he worked with industrial clients by building mood boards, scale models from Lego bricks, mock-ups from foam, and prototypes at many levels of precision, and studied interfaces with paper prototypes and multimedia models.

The first major publication of the program was the book *Empathic Design* (Koskinen, Battarbee, and Mattelmäki 2003), which renamed the group. In this book, Katja Battarbee and I defined empathy through a quote from *The Cambridge Dictionary of Philosophy*:

> *The Cambridge Dictionary of Philosophy* gives three basic meanings to "empathy." First, it denotes automatic mimicry of expressions or manifestations of emotions. Secondly, it may mean mimicry of gaze (i.e., transfer of attention from the other's response to its cause). Thirdly, it denotes role taking, which reconstructs in the imagination aspects of the other's situation as the other "perceives" it. When we talk about empathy, we refer to the third of these meanings: "empathy" is an imaginative projection into another person's situation. It represents an attempt to capture its emotional and motivational qualities. The key to empathic design is an

understanding of how the user sees, experiences, and feels some object, environment, or service when he or she uses the object. (Koskinen and Battarbee 2003, 45)

One of the first tasks that the group faced was to better define the spectacularly abstract notion of "an imaginative projection into another person's situation." A projection like this can obviously be achieved in many ways. In the introduction, Battarbee and I claimed that empathic design is user centered rather than designer centered. We also claimed that empathic insights could be gained not only by making careful observations of people, but also by using design-based methods. We finally argued that these insights could be gained through immersion in those situations in which people have feelings.

The doyen of American anthropology, Clifford Geertz, illustrated the foundational role of interpretation with a story of two boys contracting the eyelids of their right eyes. This movement may be an involuntary twitch; it can equally well be a conspiratorial signal to a friend. The difference between a twitch and a wink is vast, but as Geertz notes, a camera cannot make this distinction. The difference seems simple, but it has massive methodic implications:

> Right down at the factual base, the hard rock, as far as there is any, of the whole enterprise [of anthropology], we are already explicating: and worse, explicating explications. Winks upon winks upon winks . . . Analysis, then, is sorting out the structures of signification . . . and determining their social ground and import. (Geertz 1973, 6–9)

This idea has radical consequences for design research. These were captured in the distinction that the Polish-British social theorist Zygmunt Bauman made between interpreters and legislators (Bauman 1992, 1–25). Experts have privileged knowledge that gives them the right to draft rules and precepts that other people must follow: knowledge is power. For interpreters, knowledge has no extralinguistic standards of correctness: it is a tool to help communication between different communities. In our introduction to *Empathic Design*, Battarbee and I followed Bauman: "designers are changing from legislators to interpreters who mediate user experiences into the design process. Instead of seeing designers as legislators who know better, it has become more common to characterize design as an interpretive profession" (Koskinen and Battarbee 2003, 40).

With this distinction, empathic design became interpretive. This shift clarified the field's approach and gave it a frame through which to view its relationship to concepts and theories, two of the key constituents of research.

Meanings as Symbolic Interaction

Over the next two years, the group gave empathic design a unique interpretation by building it on social psychological and sociological studies of
emotions. These studies had argued that emotions are cognitive, social, and
cultural constructs (more about them in chapter 2). These debates had little
direct design relevance, but their message was clear. Experience is contextual
and situated, and some of its most important aspects cannot be measured
from brain waves, skin responses, or the use of any psychological instrument.
It must be studied where it happens, and research methods cannot isolate it
from the context of action.

Realizing this, the group approached experience in two ways. One impetus came from ethnomethodology. In 1999, Esko Kurvinen joined a research
project to study the future of digital imaging. Soon afterward, he read a paper
by Elinor Ochs, an anthropologist (Ochs, Jacoby, and Gonzales 1994). The
paper was about how physicists talk about images. It was a detailed study of
how a group of physicists interpreted a scientific graph turn by turn in conversation. Kurvinen had felt that methods like cultural probes were missing
rigor and detail. Ochs's conversation analysis and Harold Garfinkel's ethnomethodology (1967) gave these important elements to him.

Another impetus came from symbolic interactionism. Battarbee took her
cue from pragmatism, but she saw it in sociological rather than philosophical
terms. Her theorist of choice was Herbert Blumer, the father of the concept of
symbolic interactionism. It was easier to learn than ethnomethodology, and
maybe for that reason, it became the foundation of the group's work over
the next few years. Blumer's classic paper, which defined the term "symbolic
interactionism," described three premises that form its core:

> Symbolic interactionism rests in the last analysis on three simple premises. *The first
> premise* is that human beings act toward things on the basis of the meanings that the
> things have for them. Such things include everything that the human being may
> note in his words—physical objects, such as trees or chairs; other human beings,
> such as a mother or a store clerk; categories of human beings, such as friends or
> enemies; institutions, such as a school or a government; guiding ideals, such as indi-
> vidual independence or honesty; activities of others, such as their commands or
> requests; and such situations as an individual encounters in his daily life. *The second
> premise* is that the meaning of such things is derived from, or arises out of, the social
> interaction that one has with one's fellows. *The third premise* is that these meanings
> are handled in, and modified through, an interpretative process used by the person
> in dealing with the things he encounters. (Blumer 1969, 2, emphasis added)

People define situations and their lines of action by giving meanings to them. A person may live his or her whole life in Paris without paying any attention to fashion; another may live in Manhattan without ever visiting art galleries; a third may live in Rio de Janeiro and avoid its famed beaches, carnival parties, and soccer rivalries. Blumer also taught the group to see concepts as sensitizing devices that help inductive analysis rather than precise carriers of scientific meaning (Blumer 1954; Koskinen 2003).

In putting people and their meanings in the driver's seat, Blumer's premises brushed aside the idea that some abstract, unconscious force like a need, a preference, or social pressure drives society. Realizing this had remarkable implications for empathic design. This breakthrough led the group to start designing from people and their meanings: the main challenge in the early phases of design must be people and the meanings that they use to define situations. It also freed designers from complicated theorizing: they did not need a theory explicating what kind of deeper, invisible reality is truly behind what people say and do. Finally, it defined a site for studying experience: they knew that almost any trivial thing in everyday life could be a potential target for design. Their framework had to be flexible enough to be credible.

Blumer furthermore provided a way to understand technology and the material world. These suggest ways of acting, but people can always take other routes. Algorithms can suggest to us which books to buy, but we may not follow those suggestions. A table gives us an opportunity to have dinner, but we may also use it to play table tennis. His approach decoupled technology and human activities and turned their relationship into a question. When people encounter modern technologies or other design objects, they pay attention to them and discuss them. A few weeks later, these technologies and objects have become normalized and slip out of our attention. Many technologies are like this; they set the stage for action. If everything works well, people do not pay attention to them for long. When an object breaks and disrupts habitual lines of action, though, it becomes an object of attention, but this is special. It is unusual that people build their lives and identities around things, but there are exceptions, like collectors and recycling engineers.

During the growth years of the web and mobile telephony, the idea that people could freely choose their identities was popular, but it also faced ethnographic criticisms (Turkle 1995; Miller and Slater 2000). The question of whether people are free to choose their lines of action, or whether their choice is constrained by structural forces, intrigued Jung-Joo Lee, a talented designer who joined the group in 2006. The Blumerian vision of human action as a function of almost freely chosen lines of action did not ring true for Lee. She thought that although our identities are negotiable, people in rural Indiana or Missouri tend have much less freedom in choosing their identities than in Blumer's mushrooming, cosmopolitan Chicago, where they could choose what they wanted to be. To study what makes identities and roles stick, she investigated structural symbolic interactionists who had argued that social controls keep most of us in line (e.g., McCall and Simmons 1978; Stryker 1980, 1986).

Another researcher interested in the balance between free choice and structural grounds of experience was Krista Kosonen (2018), who studied ways in which narratives shape designers' selves through George Herbert Mead's (1934) classification of the self into the self, significant others, and generalized others, and through his distinction between the indeterminate "I" and the socially determinate "me." She came to see the former as more meaningful, perhaps because she was dealing with design, a creative and intellectually rigorous activity that requires playfulness more than respecting those deep and unconscious structures of society that keep us from causing problems. Similarly for most members of the group, Blumer's indeterminate version of human action rang truer than the structural versions Lee had explored.

Empathic Research Program

Building on an interpretive foundation, the group has by now built a large, successful, and sustainable research program. Blumer's beliefs gave it a foundation that it could use to approach the human side of design, but the beliefs did not specify how to do design research. The program was described in 2014 (Mattelmäki, Vaajakallio, and Koskinen 2014), but the main parts of the story are worth reiterating here. That paper described the program in terms of a dialectic between the core beliefs from Blumer and problem shifts. This interpretation built on the writings of the Hungarian philosopher of science Imre Lakatos (1970).

Figure 1.1 describes the advances of the program by mapping its key projects, publications, and doctoral theses with shadows and putting the program in a larger context. **Shadow 1**: As this chapter has shown, the precursor to the group's work was usability research. **Shadow 2**: The first few years of the program focused on finding ways to explore and study user experiences. This interest was a response to calls from the information technology industry and to the emerging literature about the concept. **Shadow 3**: From methods, the main problem shifted to finding ways to communicate with tools better than presentation and authored reports that did not convey their findings efficiently to stakeholders in industry. The answer to the question of how to communicate better was participation: bringing stakeholders into the design process and turning them into codesigners. **Shadow 4**: The next shift of the program was to developing design-based methods to support the creation of radical innovations. These methods came from scriptwriting, scenography, and exhibition design, among other areas. **Shadow 5**: Finally, the program developed methodologically and technologically by finding ways to integrate prototyping into its interpretive framework. **Shadow 6**: This shadow describes a few reflective papers that provide a pathway from an empathic to an interpretive framework and work that the program has inspired.

One message behind figure 1.1 is that the program has progressed dramatically. It can respond to many kinds of problems that would have been difficult to answer only ten years ago. Its methodical palette is larger, the topical scope wider, and the theoretical resources deeper. An important implication of this progress is that the group has been able to collaborate with many types of new stakeholders. These include industrial designers, interaction designers, and many engineering specialties. The group has also collaborated with several unusual stakeholders, including graphic and furniture designers, craft designers, feminists, and artists. It has also collaborated with sustainable designers and scientists from microbiology to chemistry. Some of its collaborations have been with mass media and politicians up to the level of lawmakers. These interactions have enriched the program and connected it to its social surroundings in ways that traditional design disciplines have seldom been able to do before.

Rectangular shapes around the six shadows show the main social and technological trends that have shaped the program. Its early years were driven mostly by the concerns of the emerging information technology industries: the web, mobile phones, games, social media, and other ubiquitous technology (Weiser 1991), and slightly later, the Internet of Things. Information technology has continued its sway on the program, but as the figure shows, other factors have also driven the program, especially since the beginning of the 2010s. These include normative discourse that had driven several researchers to design for the public sector and for "invisible"

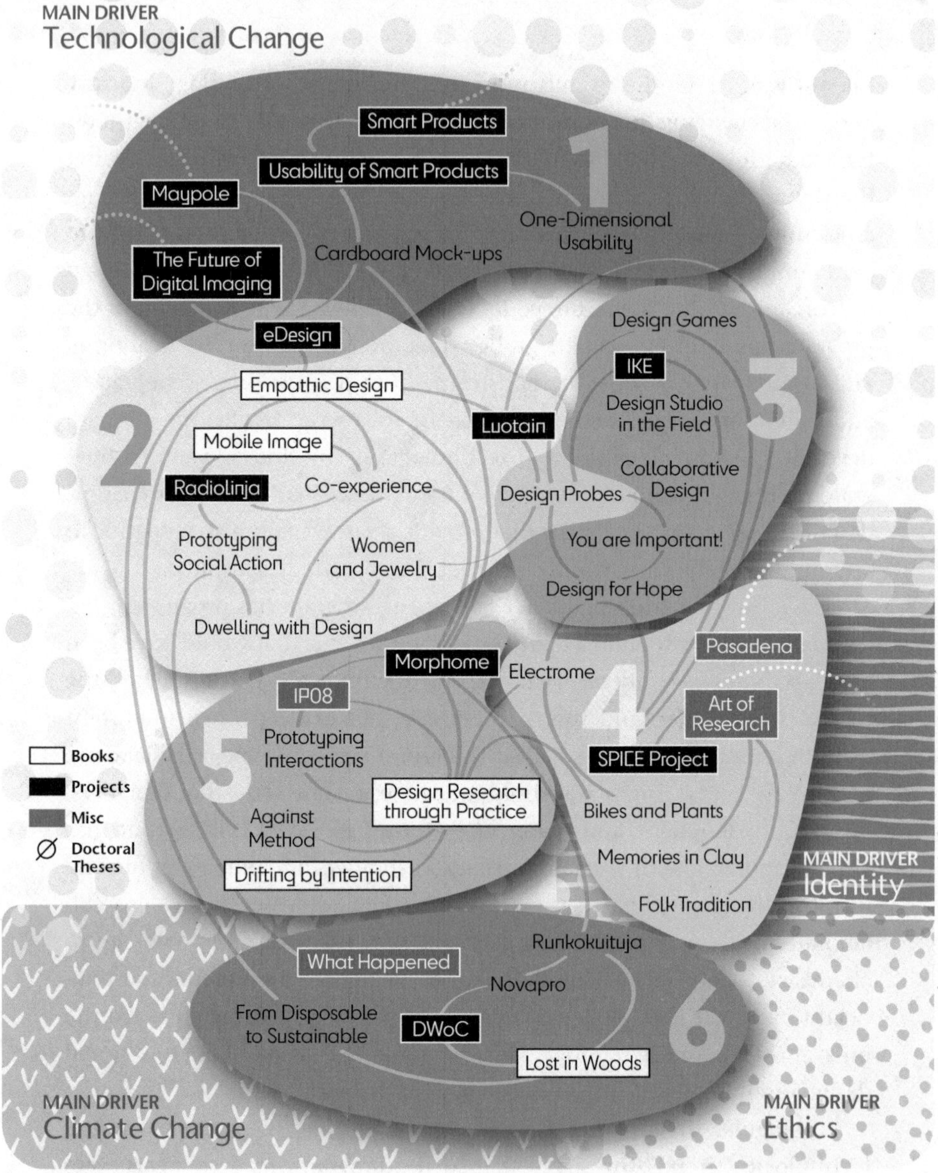

MAIN DRIVER
Technological Change

Smart Products

Usability of Smart Products

Maypole

One-Dimensional
Usability

The Future of
Digital Imaging

Cardboard Mock-ups

1

eDesign

Design Games

IKE

Empathic Design

Design Studio
in the Field

Mobile Image

Luotain

3

Radiolinja

Co-experience

Collaborative
Design

Design Probes

Prototyping
Social Action

Women
and Jewelry

You are Important!

Design for Hope

Dwelling with Design

2

Morphome

Electrome

Pasadena

IP08

Art of
Research

Prototyping
Interactions

SPICE Project

Design Research
through Practice

Bikes and Plants

5

Against
Method

Memories in Clay

Drifting by Intention

Folk Tradition

4

MAIN DRIVER
Identity

Runkokuituja

What Happened

Novapro

From Disposable
to Sustainable

DWoC

Lost in Woods

6

MAIN DRIVER
Climate Change

MAIN DRIVER
Ethics

□ Books
■ Projects
▨ Misc
∅ Doctoral
Theses

Figure 1.1
Empathic research program. Usability studies (shadow 1) led to user experience methods (shadow 2), codesign (shadow 3), explorations of design-based methods (shadow 4), and constructive methodology (shadow 5). The last shadow (shadow 6) describes how the program has responded to recent challenges. Some work in shadow 6 has gone beyond empathy, as chapter 6 will show. (Cf. Mattelmäki, Vaajakallio, and Koskinen 2014; Lee, Whalen, and Koskinen 2021.) (Picture credit: Ilpo Koskinen.)

populations (A. Júdice 2014, 9–10); discourse about human identity—especially gender, ethnicity, and tradition—that has motivated several studies to explore how people construe images of their selves, and how design could participate in this process; and climate change, which has animated research in sustainability in the 2010s.

The shadows and the rectangulars behind them require a few caveats. First, they are not based on real research. They are but broad brushstrokes that show only the most important proximate things that have shaped the program. Many other influences shape design as well, ranging from politics, industrial fads, new materials, and fiscal policies behind economic change. Second, behind many of these broader forces are antecedents that explain their rise, including the reasons for the rise of information technology and the scientific research behind a growing awareness of climate change. These were filtered into the program through issues described by the shapes. Third, these shapes are glosses: behind them are nuanced trends in areas like information technology and identity politics. For example, regarding the latter, the most relevant question for the group around 2005 was gender, but around 2015, the main topics were multiculturalism and immigrants. Fourth, the shapes do not suggest a causal path from environment to the program. The correlation between the program and the environment is "ecological" (Robinson 1950), in that its effect cannot be mapped directly onto individual pieces of research. Instead, they describe the matrix in which the program has evolved and to which it has contributed. A good analogy is on the level of a country: we may find a correlation saying that Germans are more formal dressers than Californians, but this does not mean that every individual in Germany dresses more formally than every Californian. Nationality cannot rule out individual differences. Similarly, the forces described in the rectangulars have shaped the program, but how it has actually evolved has depended on many types of choices made by individual researchers.

Having said this, it is fair to say that information technology, identity politics, and climate change have generated new research problems for the group. They have generated problems that industry and governments have responded to, and these responses have generated problems for designers. Rough as it is, figure 1.1 helps to capture an observation about the changing nature of the proximate drivers of the program; however, the basis of the program is much more complicated today than it was in its early stages.

Overview

The group's problem shifts supply the organization for this book. This chapter opened with a panorama of change in design research. The change started at the end of the 1990s, when researchers in several communities started to articulate how design itself could be turned into a research instrument through a series of usability studies. During the first few years of the shift, the line between research and practice was often blurred, but within a few years, research started to gain a separate intellectual foundation that helped researchers to better define their approach. In Helsinki, the foundation came to be symbolic interactionism. The interpretive foundation has been a source of a good deal of wealth, but it has introduced problems as well. The group has responded to these problems step by step. Building on earlier studies, it has created solutions to research problems. After finding these solutions, the program has shifted to other problems. The structure of this book follows the course of these shifts.

Chapter 1: Interpretive Turn in Design Research (shadow 1 in figure 1.1). This chapter has painted a broad picture of design research and its changes in the 1990s and the 2000s and told the early history of the main case of this book, the empathic group. The story started with Turkka Keinonen's *eDesign* project, which focused on the emotional aspects of using products. At the base of the program is a set of interpretive beliefs. From its interpretive foundation, empathic design has developed into a significant design research program. The foundation has given the program a focus and an identity. It has been guiding work for years, even though the program has changed course several times. As this book shows, the program still has its foundation in studies conducted from 1995 to 2001, but its focus has changed several times since.

Chapter 2: Making Sense of User Experience (shadow 2 in figure 1.1). The first focus of the program was in finding ways to study emotions and user experience. This was the focus of Katja Battarbee's and Tuuli Mattelmäki's empathic methods and Esko Kurvinen's and my mobile multimedia studies. Some of these studies appeared in *Empathic Design* (2003), the first main publication of the program. Key researchers were Tuuli Mattelmäki, Katja Battarbee, and Esko Kurvinen. A few years later, other key researchers included Jung-Joo Lee, Salu Ylirisku, and Yiying Wu. Key projects were *eDesign* (1998–2001) and *The Future of Digital Imaging* (1998–1999). Later, the program explored the uses of design in everyday life, with the working titles of the *Domestication of Design* project by Heidi Paavilainen (née Grönman) and Petra Ahde-Deal and the *Domestication of Ergonomics* project by Virve Peteri.

Chapter 3: Codesign and Commitment (shadow 3 in figure 1.1). The first major problem shift came from industry. Design briefs began to change from products and interactions to systems and services at around 2003. The question of who was a user and who a designer was no longer clear. In response, empathic design shifted from user-centered design to codesign, where people express their experiences in the design process. The new focus was how to communicate user experience into organizations and networks to make it effective. Key researchers were Mattelmäki, Battarbee, Katja Soini (née Virtanen), Kirsikka Vaajakallio, and Andrea and Marcelo Júdice. Key publications were Mattelmäki (2006), Vaajakallio (2012), Soini (2015), and Hakio and Mattelmäki (2011). Key projects included *Luotain*, a study of how cultural probes work in a company context (2002–2006); *Active@work* (2004–2006), a social innovation project that looked at aging workers and their well-being (Mattelmäki and Hakio 2006); *IKE project* (2003–2004), which explored housing renovation; and *Vila Rosario project* (2004–2010), which studied multi-resistant tuberculosis in Brazil.

Chapter 4: Interpretation and Radical Innovation (shadow 4 in figure 1.1). By about 2007, the empathic group was able to do research on user experience in ways that made sense to practitioners. It was also able to communicate insights to companies. The next problem shift was to finding ways to avoid an "empathy trap." At worst, empathic methods can give credence to folk opinions, which may lead to dull designs close to everyday experience. This resonated well with a message from the environment. At the end of the decade, there were several calls for radical innovation (Verganti 2009), which led researchers to find ways to let more design imagination into their work (Sterling 2005; Blythe 2014). For example, they have explored design games, surreal user research techniques, script writing with movie makers, and games. Key researchers included Mattelmäki, Vaajakallio, Andrea and Marcelo Júdice, and Sandra Viña. Key projects were *eXtreme Design* (2008–2020) and *SPICE project* (2009–2012), the first focusing on service innovation and the second on urban space around the extension of a metro line.

Chapter 5: Interpretation and the Constructive Turn (shadow 5 in figure 1.1). The next problem shift was based on a tension between the group's interpretive principles and design as a world-making discipline. As the word "empathy" suggests, the main concern of empathic design is understanding human experience. Design, however, is also an occupation that conceives and creates things that do not currently exist. The group tackled this challenge by integrating prototyping into its methodic suite. The most important study that set the tone for this part of the program was the *Morphome project* (2002–2005). The key researchers were Jussi Mikkonen, Jung-Joo Lee, and myself. The key publications were IP08 (2009), Mikkonen (2016), and Koskinen and colleagues (2011). However, these prototypes were not usually polished, their technology was at the lower end, and they were treated as means rather than ends and not always even reported in the literature. In social design, the program created prototypes and produced them (A. Júdice 2014; M. Júdice 2014).

Chapter 6: Interpretive Design Research and Beyond (shadow 6 in figure 1.1). This chapter concludes the story of interpretive design research. It describes a framework of four sensitivities: sensitivity to humans, techniques, collaboration, and design. It argues that the true meaning of the story of empathic design is that it highlights the power of interpretive ideas in design. The group's method proves that an interpretive approach to design research is a legitimate addition to scientific and artistic approaches, and this alternative also has a sounding board among the wider world of design research. The second half of chapter 6 zooms out from the empathic program to study the wider relevance of the group's experience by studying the program's impact on research communities that have moved on from technological topics that originally inspired the empathic group. The purpose of this section is to study how an interpretive framework works with topics for which it has not been designed. The book finishes with a plea for the continuing relevance of an interpretive approach to design research.

Smart Products

Usability of Smart Products

Maypole

One-Dimensional
Usability

The Future of
Digital Imaging

Cardboard Mock-ups

eDesign

Design Game

IKE

Empathic Design

Design Studio
in the Field

Mobile Image

Luotain

Collaborative
Design

Radiolinja

Co-experience

Design Probes

You are Important!

Prototyping
Social Action

Women
and Jewelry

Design for Hope

Dwelling with Design

Pasadena

Morphome

Electome

Art of
Research

IP08

SMILE Project

Prototyping
Interactions

Design Research
through Practice

Isties and Plants

Against Method

Memories in Clay

Drifting by Intention

Folk Tradition

Runkoluutuja

What Happened

Novapro

From Disposable
to Sustainable

DWoC

Lost in Woods

2 Making Sense of User Experience

The first main problem that the group faced was how to study user experience. User experience was an intriguing concept, but it was unclear what would be the best way to study it from a design perspective.

The empathic group studied the methods first in *Empathic Design* (Koskinen, Battarbee, and Mattelmäki 2003), which articulated the beliefs that inspired the group for the next decade. The methods were visual and tactile, cheap and low-tech, interpretive, playful and fun, tested, and targeted at the fuzzy front end of product development.

The group then explored cultural probes and ethnography in the *Luotain* project and technical platforms in *Mobile Image* and *Radiolinja*. Its main contributions were the notion of co-experience, which saw user experience as a social process, and a methodology for prototyping social action.

New researchers joined the group in 2004–2005 to study experience in everyday life rather than in a use situation. *Dwelling with Design* (Paavilainen 2014) explored how design shaped the home, and *Women and Jewelry* (Ahde-Deal 2013) how jewelry can carry experiences over generations.

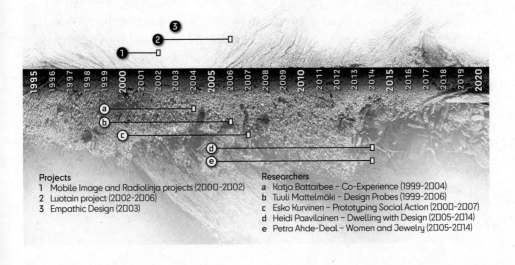

Projects
1 Mobile Image and Radiolinja projects (2000-2002)
2 Luotain project (2002-2006)
3 Empathic Design (2003)

Researchers
a Katja Battarbee – Co-Experience (1999-2004)
b Tuuli Mattelmäki – Design Probes (1999-2006)
c Esko Kurvinen – Prototyping Social Action (2000-2007)
d Heidi Paavilainen – Dwelling with Design (2005-2014)
e Petra Ahde-Deal – Women and Jewelry (2005-2014)

Before the end of the 1990s, the key words that design researchers were familiar with came from neighboring disciplines. Ergonomics talked about "stress" and "load," semioticians about "meaning," management about "leadership," marketing about "corporate image" and "branding," and cognitive psychology about "knowledge" and "information processing." As chapter 1 showed, however, design increasingly started to converge around the concept of "user experience" by the end of the century. With this term, design got a footing in a philosophical concept that was flexible enough to be interesting for designers. It did not have one definition; industry used it, and it was easy to do so without a sophisticated theoretical background. Its scope was vast: we are always experiencing something. By 2002, user experience had become the mainstay of design research. It was a powerful tool for designers who wanted to create pleasant, delightful, and attractive products and interfaces.

Yet there was a problem. As soon as user experience became one of the primary concepts in design, researchers had to find ways to study it. In Helsinki, Turkka Keinonen and Tuuli Mattelmäki were particularly frustrated with usability studies that did not support creativity, which was one of the reasons behind the latter's decision to embark on doctoral studies. Her motivation was clear, but she did not have a vision of what design-based methods could be like. Several candidate answers entered the stage when she started her work. At one extreme were frameworks from cognitive psychology (Hassenzahl 2004). In *Emotional Design: Why We Love (or Hate) Everyday Things*, Don Norman (1998) argued that designers need to stress emotion rather than cognition. At the other extreme were methods from industry, including wire framing and sketching, ethnography, experience prototyping, and cultural probes (Shedroff 1991; Beyer and Holtzblatt 1998; Sanders 1994; Buchenau and Fulton Suri 2000; Dunne and Gaver 2001). These methods introduced unusual theoretical references: for example, probes introduced surreal and situationist references into design research. The array was dizzying.

Although there were several precedents for it, a consistent interpretive approach to research was missing. Participatory designers had learned in the 1970s that mock-ups work better in communication than technical concepts (Ehn and Kyng 1991), and there was a body of interpretive and critical literature in information systems (Hirschheim and Klein 1989;

Mumford et al. 1985). Contextual inquiry captured experiences in task-based work environments (Beyer and Holtzblatt 1998); researchers at Intel used immersive role-plays (Salvador and Howells 1998), SonicRim had developed context mapping and Velcro modeling as research tools (Sanders 2000), and IDEO's Marion Buchenau and Jane Fulton Suri (2000, 427) introduced bodystorming, defined as "physically situated brainstorming," and experience prototyping to design research. The ingredients were there, but the recipe was missing.

This chapter tells the story of how the empathic group defined its methodic approach to experience. It happened in a few main phases. Simo Säde (2001) had explored three-dimensional (3D) mock-ups, puzzle interviews, and prototypes as research methods in his thesis, which focused on usability. Katja Battarbee, Tuuli Mattelmäki, and Esko Kurvinen morphed Säde's interest in usability into methods for studying user experience. Battarbee had conducted contextual studies with existing products in the 1990s and continued her work in the early 2000s, Kurvinen was interested in social action in multimedia phones, and Mattelmäki focused on turning cultural probes into interpretive instruments. Several master's theses and classroom projects had created methods that focused on underexplored aspects of experience. These studies found their first articulation in the *eDesign* project and in the book *Empathic Design* (Koskinen et al. 2003). The next generation of researchers pushed these studies from looking at use situations to larger contexts of life under the rubric of domestication.

From *eDesign* to *Empathic Design*

The empathic approach was first formulated in *eDesign—Design for Emotional Experience in Product Use* (1999–2001). Its original goal was to find ways to measure emotions with sensors embedded in mobile phones and other small accessories. During the project, it became clear that standard measurements like heartbeat and galvanic skin response could give only very crude and imprecise measures of emotions. The focus was on sensor technology and algorithms (Picard 1997). From my perspective, the premise behind the idea of measuring emotions with sensors was theoretically and conceptually suspicious. It built on behavioristic assumptions that painted humans as somehow driven by electrical and chemical reactions in the body. For me, this was obviously true, but I thought that emotions were also—and primarily—social, cultural, and tactical phenomena.

The shift to a social interpretation of emotions became clear in the project's steering group meeting in 2001. Panu Korhonen was a mathematician working for Nokia Corporation and the chair of the steering group. In the meeting, he had an exchange with me. I had argued that it was necessary to study emotions by interpreting how people experience them before making any design decisions. Korhonen quipped that this was something that designers would never do: in his experience, designers prefer to stay in the studio instead of studying people seriously. I responded that the project had changed philosophy. Both of us laughed, and the meeting moved to other topics.

The message from this exchange became clear to me over the next few weeks. I had studied social psychology years earlier at Indiana University, and this experience had convinced me of the dual nature of emotions. There was no doubt about the fact that human emotions have a biological and electrochemical basis in the body, and I believed that these reactions could be measured (at least in principle). However, except for fear and sexual arousal, which are key to survival of the species, I also believed that these electrochemical processes do not map directly onto emotional experience (Schachter and Singer 1962; Kemper 1981; Shott 1979). For example, people think and talk about their emotions, use them tactically and commercially, and organize their own activities and other events around them (Katz 1999; Hochschild 2003; Bailey 1983; Rosenberg 1990). The main gap in the design debate back then, as far as Korhonen was concerned, was how to capture the social side of emotions.

This gap was conceptual, but also methodic. My reference point was an extensive body of literature about emotions in psychology, social psychology, and sociology that I had read in Indiana. This literature convincingly argued that emotions have a social dimension, but there was little evidence that this perspective works in design literature. The first attempt to start creating this proof was the book *Empathic Design* (Koskinen et al. 2003). As its main editor, I saw fascinating design research around me, but I also thought that its theoretical base had to become stronger and more coherent. I also knew the mentality of the researchers around him. In my eyes, they were clearly interpretive. They all had a bottom-up approach to research, and they avoided statistical methods. Their colleagues in industry also preferred close observation rather than statistics. *Empathic Design* was a chance to show that

design research can be built on an interpretive basis without massive theoretical baggage, and research can focus on design methods. When researching for the book, it quickly dawned on me that those similar ideas were in the
air in design in several places. These included John Dewey's (1980) pragmatism and Jerome Bruner's theory of storytelling (Bruner 1990) in interaction
design, and Martin Heidegger's notion of being-in-the-world (2010) in art
and the emerging field of craft research in Lund, Sweden, and Nottingham,
England (Gedenryd 1998; Niedderer 2004). All these could be indications
that design researchers had to revise their theory of emotions.

Building on my knowledge of social psychology, I started seeing design
methods as research tools into emotions and experience. My reasoning
was ethnomethodological: if good designers have methods that work in
practice, design researchers should base their craft on these same methods. These methods also had other qualities that made them attractive. In
the preface of *Empathic Design*, I noted that the work leading to the book
was firmly rooted in user-centered design and described a shift from ethnographic to design-based methods. As I also wrote, empathic designers had
consistently followed a few principles. Looking at the collection of papers
in the book, I noted that the methods were *user-centered*, first and foremost.
They were also:

- *Visual and tactile*, providing designers with inspiration, not just data.

- *Deliberately cheap and low tech*, and as such, easy to adopt in the real
  world, where money is scarce.

- *Interpretive*. To be able to design effectively, designers need to understand
  how people understand themselves.

- *Playful and fun*. When exploring new ideas, users are almost invariably
  asked to imagine and dream in a future world created by designers. To be
  rewarding, such exercises must be fun.

- *Tested in reality*. We report cases from real product and concept development because we believe that this is the best way to make sure that our
  proposed methods work where they should: at the front line of imagination in the corporate reality.

- *Targeted at the fuzzy front end*, as Jonathan Cagan and Craig Vogel (2002)
  from Carnegie Mellon University recently called the early phases of
  product development (Koskinen et al. 2003, 7–8).

The introduction of the book distinguished preexperience from experience and postexperience and set out to map methods used by the local design community for capturing experience over its life cycle. In the book, two talented industrial designers, Sauli Suomela and Arni Aromaa, described how they designed a smart bathroom for Oras, a leading faucet and bathroom equipment manufacturer, and Jane Fulton Suri described her encounters with universal design and bodystorming. Esko Kurvinen showed how to study emotions in mobile multimedia, and with me, he wrote about an experience prototype that we had made to study the integration of mobile phones to web-based photo albums, a distant dream back then. Meri Laine (2001) described how she had used objects, photographs, and interviews to study how people experience carrying things, and Tuuli Mattelmäki formulated her interpretive approach to cultural probes. Katja Battarbee described how storytelling could make sense of user experience, and Alison Black wrote about her experience as a cognitive psychologist at IDEO.

In hindsight, the main contribution of the book was its interpretive approach, not any particular method. As the years went on, I stressed that the methods as such are far less important than the framework. My analogy was statistics, in which knowledge of, say, the basics of linear models helps us to understand the principles behind percentages and cross-tabulations, as well as two-phase regression, logit models, and structural equations. I reasoned that this is also the way forward in design research. It is easy to list several hundred design methods from about ten well-known books about design methods and get lost in semantics. If you see the big picture behind them, though, you can be flexible with methods and avoid being confused by the sheer numbers.

Observing Technologies That Do Not Exist: *Prototyping Social*

Empathic Design left a twofold methodic legacy. On the one hand, papers by Kurvinen and me built on conversation analysis and ethnomethodology, both of which are philosophically consistent with phenomenology. On the other hand, Mattelmäki formulated an interpretive approach to cultural probes that had been introduced to design literature a few years earlier by Bill Gaver, Elena Pacenti, and Anthony Dunne (1999). The former approach was built on ethnography, and its goal was to describe the human world for design; the other was built on mainly surreal and situationist sources, and its goal was to inspire designers. These approaches lived side by side in the research community of the design school for a few years.

In 2003, when *Empathic Design* was published, both approaches seemed equal in strength, and both were attracting a following at international design research conferences like Designing Interactive Systems and Designing Pleasurable Products and Interfaces. In a few years, however, it became clear that Mattelmäki's approach was more attractive to design students, who had a hard time believing in the benefits of description. They valued evocative, curiosity-rousing methods rather than precision and descriptive power.

Kurvinen's methods were working well in industrial work. He is an industrial designer who had been hired to work on the *Future of Digital Imaging* project in early 1999. This project's goal was to study the future, which is omnirelevant for designers whose discipline is interested in creating things that do not exist (Alexander 1964; Borgmann 1995; Cross 2001). In search of a method, Kurvinen started with a literature review and found that the main future-oriented design approach of the day was trend analysis, cool hunting, and street ethnography, which worked well in fashion but far less well in high-tech enterprises (Vejlgaard 2008; Gladwell 1997; Polhemus 1994). Future studies, for their part, usually centered around statistical forecasting and worked on central tendencies and dispersion measures that were difficult to translate into design proposals. Artistic imagination, which was coming to design research through Anthony Dunne's doctoral thesis in London, led to interesting perspectives and charming one-off designs but, again, it was not convincing in the hard-nosed world of industrial design (Dunne 2005; Dunne and Raby 2001).

Kurvinen, by contrast, wanted to ground his research in empirical detail and developed a method to study nonexisting human interaction empirically for design purposes. These studies addressed the fact that design is a world-making discipline. With Turo-Kimmo Lehtonen, a sociologist who had joined the project in 1999, Kurvinen and I recruited three groups of friends and gave every participant a Nokia 9110 Communicator (the first true smart phone) and a digital camera by Casio. The camera was able to beam images into the communicator with an infrared link and send these images as email attachments. At least in the Western markets, the Communicator was the first phone that was able to do this.

The system was slow and the quality of images low by present-day standards, but this piece of technology offered a research proxy for Kurvinen. He built a system in which users were able to capture and send their messages to their friends on the fly, read them, and respond to them. He could follow these multimedia conversations via his email. With me, he examined how to study interactions message by message. The basic idea behind this came from the father of ethnomethodology, Harold Garfinkel (1967), and the father of conversation analysis, Harvey Sacks (Sacks and Schegloff 1974): when one person sees something interesting and captures it, he or she can share it in an image, pick up elements from it in text, and send this message to others. When others see it, they make sense of it and may respond to it. The responses in turn show the sender of the original message how they understood it.

In this framework, things like emotions become interactive and observable things: if I take a photo of a pulled pork sandwich and say that it is delicious, the responses can vary from "Yummy" to "You are destroying the Earth again." A heated response to the latter reply can lead to an accusation about "being a m@r@n." This response may escalate into a fight about character, but it can also be a tease, with a statement like "Yep, and proud of it, like you." The next message, in turn, depends on this response and shapes the conversation again. "Don't play with me, I am serious" starts an argument. "I salute you," with a smile and wink, turns it into a joke. And so on.

Kurvinen studied mobile multimedia in two platforms. *Mobile Image* was a university-built system that used existing technology—a platform that let participants capture and share messages with a camera and a mobile phone. Kurvinen captured all the messages in the inbox of his email unless the senders marked them as secret or asked him to destroy them. *Radiolinja* was a technology pilot project by Radiolinja, the mobile arm of the former Helsinki Telephone Company (now Elisa). Kurvinen participated in the pilot as an external consultant. With the company, he built a web link that showed messages between the pilot's participants in real time. A research assistant captured and saved them in Adobe Photoshop once a day unless the participants of the study destroyed them. *Radiolinja* produced a data set of over 4,000 unique messages.

The high point of Kurvinen's research was *Prototyping Social Action* (2007). He built on Battarbee (2004) and presented a paradigm for studying experiences as they happen within a technology prototype. The paradigm had four main components:

- *Naturalistic research design and methods.* People are the authors of their own experiences. They are involved as creative actors who can and will engage with available products that support them in interests, social interactions, and experiences that they find meaningful. Data from people must be gathered and treated using empirical and up-to-date research methods.

- *Openness.* The prototype should not be thought of as a laboratory experiment. The designer's task is to observe and interpret how people use and explore the technology, not to force them to use it in predefined ways.

- *A sufficient time span.* The prototype usage ought to be observed for a long enough time (typically for a few weeks at least) since it is difficult to get an idea of how people explore and redefine the technology in their actions if the study period is shorter. However, as our third example in this discussion shows, one can create platforms to see how people use the prototype using considerably shorter study periods, provided that the setting is open enough for the participants to organize their activities freely around the prototype.

- *Special attention to the sequential unfolding of events.* One needs to study the stepwise development of the social process, not simply list its outcomes. Interaction unfolds over time and has to be considered in temporal terms (Kurvinen, Battarbee, and Koskinen 2008, 49–50).

The beauty of this paradigm was that it gave a blueprint for empirically observing technologies that do not (yet) exist. The key word is "empirically": Kurvinen's studies offered a far more credible picture of the future than scenarios and other products of imagination that designers had usually been doing. For a future-oriented discipline, this is crucially important. The basic idea had been in the air for quite a while, of course. For example, the *Maypole* project had built prototypes of mobile phone cameras (Mäkelä et al. 2000), concepts like IDEO's "experience prototyping" were already bridging the gap between design and design research, and other researchers were exploring concepts for turning design artifacts into probes into the future (Crabtree 2004). Kurvinen's approach worked on a large scale, and ethnomethodology gave him sophisticated theoretical and methodic tools to study how technology may transform age-old forms of experience.

The main novelty of his paradigm was the attention to the sequential unfolding of events. The paradigm was elaborated in three books: *Mobile*

Image (Koskinen et al. 2002), the first empirical study of mobile multimedia phones; *Mobile Multimedia in Action* (Koskinen 2007), targeted at communication scholars; and *Prototyping Social Action,* written for design researchers (Kurvinen 2007). Building on ethnomethodology and conversation analysis, these books painted a picture of mobile multimedia as social phenomena that develop over time in ordinary, mundane activities. Kurvinen continued his studies with software engineers, supervised master's theses that applied these methods in field settings, and later inspired Jung-Joo Lee, who became the final ethnomethodologist in the empathic group. By 2012, both Kurvinen and I had left the group, and after Lee moved to Singapore in 2014, the group did not have the theoretical knowledge to keep the ethnomethodology alive.

Kurvinen's paradigm lived on in the group after he left, however. How his approach related to design was unclear for a while, but the issue became clear in two conversations with professional social scientists.

First came a series of conversations between Kurvinen, me, and Ilkka Arminen, now a professor of sociology at the University of Helsinki and my old friend from our student days (Arminen 2004). Arminen was fascinated about the ethnomethodological approach of *Mobile Image* (Koskinen et al. 2002). The conversation that we all had led us to realize that although our methods were similar, our perspective on time was profoundly different. The series took place in 2002. As a sociologist, Arminen wanted to plan a study that would have started around 2003. Research would take a year. Getting to major publications would take another two to three years—that is, the first genuinely good publications would not appear until 2006. But in 2006, designers would be interested in 2008. The year 2003 would be history for them, so there was a five-year gap that made genuine collaboration difficult. In 2008, Tuuli Mattelmäki had a similar conversation with Pasi Mäenpää, another sociologist from the University of Helsinki who participated in the *SPICE* project (see chapter 4). Mäenpää told her that he could only study reality. The implication was that he would have to wait until the line was ready before he could study it. For designers, this is impossible. They live in possibilities rather than facts. Again, the issue was the gap between description and explanation, which constitute the goal of scientists, and inspiration, which was the goal of design researchers.

The Path to Probing

When *Empathic Design* was published, the group was well familiar with an emerging trend in design research—a move toward ethnography. Ethnographic research that was particularly prominent in the United States. IDEO encouraged designers to "shadow" users, the Palo Alto Research Center had world-leading ethnomethologists, Rick Robinson's E-Lab and Intel both hired ethnographers (Wasson 2000), and Karen Holtzblatt's *InContext* turned contextual inquiry into a business success. These approaches had roots in a variety of areas: design practice, ethnomethodology, symbolic anthropology, and participatory design, but their common thrust was context. They encouraged designers to leave the studio to study technology in those contexts in which people use it. Closer to home, fieldwork was done by participatory designers (Ehn 1988), activity theorists (Kuutti 1996; Nardi 1996), and design anthropologists in Denmark.

Just like Kurvinen's ethnomethodology, ethnographic methods failed to strike a chord among the rest of the empathic group. They were too far from designers' experiences to be exciting, and too focused on observation and description to answer concrete design problems. For their part, design students learned ethnographic methods if they had to, but their response to them was far from enthusiastic.

The methodic line that became popular came from Mattelmäki. She built on cultural probes in Dunne and Gaver's *Presence Project* (2001), following the *Maypole* project. She had also heard Bill Gaver's presentation in the first Design and Emotion conference in Delft, the Netherlands, in 1999. Cultural probes were things like diaries, disposable cameras, and gamelike, playful tasks that were sent to people who could use them at their leisure. They gathered samples from culture—the obvious analogy being probes into outer space or the deep sea with the purpose of gathering small samples to analyze in the laboratory. In 1999, Mattelmäki started exploring well-being and exercising for Polar Electro, a heart-rate manufacturer. She wanted to find a method that could explore people's lives broadly rather than focusing on, say, dietary habits only. Probes proved to be a practical way to explore diabetes. The results of her study were surprising, and they disrupted some of the company's dominant medical beliefs. To the chagrin of the doctors consulted by the company, her probes showed that diabetics routinely eat sweets and fatty foods.

The Polar study led her to ask whether probing, with its roots in psycho-analysis and avant-garde art, would work in industry. This question was first formulated in her chapter "Probes: Studying Experiences for Design Empathy" (Mattelmäki 2003). Her first answer was positive, but she also came to propose a change to the way in which Dunne and Gaver's *Presence Project* had described probing (2001). While for Gaver, probing was primarily about inspiration, Mattelmäki came to see it as interpretation. Unlike Gaver, she went back to people to check whether people recognized themselves from the story that she had been putting together from the probe returns.

The act of checking gave her probes an interpretive overtone. It worked against the original intention of the probes, but it gave them credibility in industry. It also filled the gap in the *Presence Project*, which had argued against the idea that design is science but had neglected interpretive approaches to research. Ever since, Mattelmäki has supervised dozens if not hundreds of student projects and master's theses, and her approach has influenced a generation of doctoral students. It was clear that the students found the probes attractive and engaging, and her thesis proved that they could work in industry.

Conceptual Pivot: Co-experience

The final discussion that came out of *Empathic Design* was conceptual. One of the editors of the book was Katja Battarbee, who soon became the most influential theorist of the group. Her work on co-experience is well known in design, human-computer interaction (HCI), and other fields, even after she withdrew from research in 2007 after moving to California and taking a job in industry. Her concept of "co-experience" invited designers to see that experience almost always takes place in a social context. If experience is separated from social context, she argued, it is simplified and misspecified significantly. Studying co-experience means looking at how people lift things out of stream of consciousness, how they keep things in focus, and how their attention shifts to other things.

The first place in which she explored experience through philosophical and sociological approached to experience was the introduction to *Empathic Design*. In conversations with me, she came to doubt psychological theories that dominated European discussions about experience back then (for other critiques, see Djajadiningrat, Overbeeke, and Wensveen 2002). Dissatisfied with these theories, she started to explore experience as a social phenomenon in context. For instance, she analyzed a case in which a young man in a baseball cap was driving in an old car to a rock festival. He took a photo of himself driving the car and described how happy he was about the festival. Recipients shared his joy. A few minutes later, he shows a speeding ticket, accompanied by a curse, which led to commiseration from his recipients. With these messages, he made his experience visible to others and gave them an opportunity to take part in it. Similar processes can be seen when, say, teenagers play video games or people go to see concerts or watch sports. "Co-experience" is an apt word to describe these moments.

The path to the concept was long but painless. Battarbee had published almost twenty papers between 1997 and 2003, but these papers were coauthored and written over the course of several projects. She had much more content than a doctoral thesis needed, but she lacked a storyline that would have tied all this work together. The turn took one conversation. She was sitting in Kipsari, a student cafe in the basement of the university with me, her doctoral advisor. During our conversation, I noted that her main interest was how people have fun together: this was the thread that kept her work going. Because she was also interested in user experience, I suggested that she could put these two interests together and study experience as a social achievement. Half-jokingly, I suggested that she should talk about "co-experience" instead of "having fun together" or "user experience," and added that this would be a novelty in the research literature, which was dominated by individualistic frameworks.

Electrified, Battarbee finished her coffee and rushed back to her office. In a few hours, she had written a short paper titled "Co-Experience—The Social User Experience" (Battarbee 2003a). She sent it to the Computer-Human Interaction (CHI) conference, and it was accepted. Then she wrote three other papers in quick succession. One elaborated on the short paper, another looked at co-experience as a process that happens turn by turn, and the third specified the relationship of co-experience to the model of user experience proposed by her friend Jodi Forlizzi (Forlizzi and Ford 2000). After these papers were published, she had a clear path to her doctorate: she could show the steps from her earlier interest in having fun to co-experience, and she could show that co-experience can in fact be turned into a design tool. Figure 2.1 illustrates some of the processes of interaction that she saw in her studies.

Figure 2.1
Co-experience: an illustration. Katja Battarbee gave the Philips in2it camera and Nintendo Game Boy Camera to children and followed what they did with it. The drawing illustrates how they dramatized situations for their friends. Here, two boys are creating a fighting scene using props like chairs. (Picture credit: Katja Battarbee.)

The preparation that made this concept possible had taken about three years. She had explored fun in two studies during the *Maypole* project. One study was a pilot that mapped how children played with the Philips in2it camera and Nintendo Game Boy Camera (Mäkelä and Battarbee 1999). Another study created an early prototype of a digital camera that could capture and share photographs over a mobile phone network. She had conducted these studies with the psychologist Anu Mäkelä. She went on an exchange to Carnegie Mellon University, where she worked with Forlizzi. She also took part in European doctoral schools and got data from Kurvinen's *Radiolinja* project, which will be described later in this chapter. After she discovered the notion of co-experience, she saw a thread running through all the earlier studies.

Finishing the thesis was straightforward after she had discovered her concept. Once she had claimed the concept as hers, the remaining task was to write an introductory essay that showed how these papers led to her invention. She also realized that her work had roots in pragmatism when she spotted a footnote to Dewey in Forlizzi and Ford's paper. From me, she also learned that symbolic interactionism had developed in Chicago simultaneously with George Herbert Mead's pragmatism. I also suggested that symbolic interactionism might give her a way to understand experience in social terms. Her exploration into sociology culminated in a paper that she coauthored with me (Battarbee and Koskinen 2004). It was reprinted in her doctoral thesis, which became the theoretical cornerstone of empathic design (Battarbee 2004).

Beyond the Definition of the Situation

Battarbee's doctoral thesis was examined in 2004. By that time, her work had exposed a major problem in user experience literature, which depended on the notion of the "situation." Her framework helped to analyze how experience happens in a social context, but it did not go beyond it. Still, it is obvious that situations arise from something. People in poor neighborhoods get in trouble with the police more often than those in rich parts of town. People with design experience spot inconsistencies in interfaces quicker than a layperson does. Some factor leads one to put more police on the streets in slum areas, buy better equipment, or learn the skills needed to navigate interfaces. The situational approaches worked well with information technology but were less helpful in making sense of how people experience clothes, furniture, jewelry, or ergonomics over long periods of time.

For me, the "something" behind situations posed a theoretical question. The agenda of the early years of the program had been based in part on ethnomethodology, but symbolic interactionism had proved to be easier to work with for most designers. Although it gave researchers tools to study experience as a social process, its key concept was the belief that "if men define situations as real, they are real in their consequences" (Thomas and Thomas 1928, 527). This idea is powerful for understanding what people do in ordinary situations, but it has less power when activities take place over a long time or in a context that most of us are barely aware of.

The group tackled these limitations through the concept of "domestication." The idea came from Roger Silverstone, who had studied the domestication of media technology in England (Silverstone 1994; Silverstone and Hirsch 1992; Silverstone, Hirsch, and Morley 2003; see also Pantzar 1996). His concept was evocative and intriguing, but it soon became clear that his framework was not specific enough for the purposes of design. To enrich the body of literature on domestication, Battarbee studied several sources with me, including Mihaly Csikszentmihalyi and Eugene Rochberg-Halton's *The Meaning of Things* (1981) and David Halle's *Inside Culture* (1993). These sources provided anecdotal evidence about a variety of objects, including stuffed animals and prints of artworks. Then he studied several precursors from within the group. One was a study of the role of design in the manufacturing industry (Järvinen and Koskinen 2001), another explored how investors understand design (Aspara 2009), and a third suggested that the fate of design depends on nondiscursive processes, including competition from other objects at home (Nieminen-Sundell and Pantzar 2003). By turning these studies into precedents, I figured that Herbert Blumer (as already mentioned in chapter 1) might be a good guide to use to make sense of the larger context around user experience because his writings gave concepts that could be used to study experience in social context directly and did not require as much sociological background knowledge as other interactionist theories. So I hired two doctoral students to study domestication from two different angles to study user experience in everyday life. These students were Heidi Paavilainen and Petra Ahde-Deal.

Dwelling with Design: Design at Home

Paavilainen wanted to learn what happens to design when it enters the home. In her doctoral thesis *Dwelling with Design* (2014), she studied seventeen households in Helsinki for about four years. During that time, she saw people moving, houses built, marriages and divorces, and children growing. Understanding this context, she assumed, is crucial for answering her research question.

She realized that with few exceptions, the ways in which design enters homes depends more on how people define their homes than on their definition of design. She identified three primary definitions of the home. When a home is a "hotel," it is a place for meals, sleeping, hygiene, entertainment, and socializing. As she noted, this means that its design is usually meant to be durable and easy to maintain, and it is weakly integrated into everyday life if it functions well, is easy to clean, and cheap to replace (Paavilainen 2014, 196). When a home is a "museum," people focus on curating the future value of objects and the space within which they are contained. They may not be personally pleasing, and they may not be display items either. The governing principle is that things must increase in value—but not necessarily monetary value—over time (Paavilainen 2014, 203–205). Finally, when a home is a "gallery," it has been built for spiritual recreation. Curators create displays of similar objects and make sure that things fit into a larger pattern or style. Objects get a prominent place, and they are surrounded by an empty, gallerylike space. Curating a space like this can be time-consuming and may involve activities like finding vintage screws for cupboards and lights, for instance.

According to Paavilainen, most homes change over time as new things are added to them. Many objects at home are also in limbo, waiting for a final decision about what they are about or for. For example, objects can be absent for quite a long time: a wall can be kept empty until the inhabitants find the right bookshelf (Paavilainen 2014, 72). The value of objects lies more in these definitions than in their design. Some people place design objects on a pedestal; others collect things like lights from the 1950s or kitchenware from the 1960s; and yet others see even valuable design objects as barely more than pieces of junk.

A novel picture of design started to emerge from her study. Design was one of the things that people pay attention to in everyday life, but it is often far from the top of the list of life's priorities. For example, there are new beginnings in life: marriages, babies, new houses, new jobs, new towns. People renovate their homes. They inherit things and property. Children leave and take things with them. There are also endings. Houses burn. Relationships grow bleak and break. Divorces break families, and they also break connections to objects. Design objects participate in these events, but they are seldom sacred. They are like props on the stage: a part of the constitution of everyday life, but often unnoticed and almost always replaceable.

Paavilainen concluded that design is only one thing that people consider when curating their homes. Some take an easy approach to curating. They buy things when they need them and pay little attention to their aesthetic or value implications or how their home looks to outsiders. Some pay slightly more attention to things they buy, but buy things to please themselves only. Yet others devote considerably more time and pay more attention to things at home, but they are confused about their decision criteria and prefer to keep things on hold until they know what they want. This state of "confused care," as Paavilainen called it, means that a home can be years in the making.

One thing that Paavilainen did not study was the relationship of the domestication process to the social and cultural background of her subjects. This was probably a result of using Helsinki as a research site. Design objects—including classic pieces of glassware and furniture—are ubiquitous in that town. This also struck Roger Silverstone, Paavilainen's mentor in London. Silverstone noted to Paavilainen that he had always seen design as an expression of a household's economic and cultural capital, which apparently was not the case in Helsinki, which had had a different path to modernity than his hometown.

Women and Jewelry: Design That Transcends Life

Dwelling with Design studied ordinary design objects like tables, chairs, vases, sofas, cutlery, phones, and television sets. However, there are far more than consumables. Some objects may even become symbols of identity and may come to carry family identities beyond the individual. Ahde-Deal was interested in what gives some objects such a transcendental quality. A goldsmith by training, she was interested in jewelry. Her interest came from a personal doubt: she found herself questioning why she was bringing small, valuable objects into a world already saturated with things. To rebuild her belief in what she was doing professionally, she wanted to study pieces that had been in a family for generations. Studying them seemed to provide an obvious way to tell her why her work mattered.

Her journey started from her master's thesis, *Bling Bling* (2006). In it, she interviewed and photographed a multiethnic group of teenagers in East Helsinki to learn how they see jewelry and other adornments in an era in which fast-fashion companies like H&M and Zara had turned jewelry into a seasonal accessory. She learned that for these girls, the meaning of jewelry did not lie in its material or its design, but rather in the relationships that the individual items represented. A piece of colored string with knots may be as meaningful for them than a golden earring that they had received from their family for confirmation or bat mitzvah. Many items were meaningless for outsiders but full of meaning for the girls who knew their story. Conspicuous jewelry bought from fast fashion stores were multiple, but also the least meaningful. The girls referred to them as "bling bling," which gave the name to her thesis.

Bling Bling became a pilot of Ahde-Deal's doctoral study, *Women and Jewelry* (2013), which explored jewelry in adulthood. It focused on those pieces of jewelry that have stayed in a family for generations. As a goldsmith, she knew there are pieces that are kept in the family for decades, or in some cases for centuries. She also knew that most of this circulation took place in the female line, and it was largely invisible to men.

Ahde-Deal collected material from three sources. She got the rights to over 400 life histories from a biography-collecting research project by the Finnish Literature Society and Kalevala-Koru, a Finnish company inspired by the national epic *Kalevala*. These stories had their share of biases, but they also gave her an image of objects over exceptionally long spans of time. She moved to Chicago and designed cultural probes to elicit stories and memories of jewelry for detailed interviews at home. She repeated the Chicago study after she moved back to Helsinki.

When she started to make sense of these materials, she realized two things. First, it was clear that the market value of a piece has little to do with its emotional value, and as a rule, the latter was more important for the women in her sample. Jewelry carried memories of family tragedies and deceased relatives, but also births and other successes in life. Women kept these memories alive in rituals, ceremonies, and rules about ownership. She heard about rings and necklaces that had to be handed down from grand-mothers to granddaughters, about rules telling what to do if there are only male heirs, and about elaborate rituals in which these vessels of meaning were to be given to younger generations. Second, it was clear that the value of jewelry is mostly social and lies in experience: there is little connection between the monetary value of a piece and its emotional value. For the most important items, women saw themselves as keepers, not owners. Their elders had entrusted them with pieces that they were to hold until they could hand them to their offspring, to connect the family over several generations. As keepers, they were only a part of a larger chain of existence, and this chain was more meaningful than themselves.

Women kept repeating to her how their jewelry carried "powers" that they had inherited and had to transfer to their offspring. The secret of these powers lay in memories and stories that gave women the power to deal with issues in their lives. These memories and stories told them how a grand-mother had survived the war, or how she had survived the death of her fiancé, who had drowned at the age of twenty. A story of how the grand-mother had survived the loss of a family member has provided solace and resources for generations in the family. If a piece had given her strength, it might do that again. In this way, these pieces became amulets that gave women power to face the difficulties of life. Some pieces of jewelry also carried secrets that were to be kept secret, hidden.

This was a fascinating and robust observation, but Ahde-Deal needed a theoretical framework to make sense of the origins of these powers. It came from Christena Nippert-Eng, a design sociologist she worked with at the Illinois Institute of Technology in Chicago (she is now at Indiana University). In Chicago, they came to see a similarity between the stories that Ahde-Deal had heard and the argument that the French sociologist Émile Durkheim had made over 100 years earlier in his classic *Elementary Forms of Religious Life* (1912/1980). Durkheim had argued that the meaning of totemic animals and other visible religious symbols of clan affiliation among Australian aboriginals lies in their social function. He observed that while the aboriginals spend most of their days in separate, small groups, occasionally the clan comes together for religious festivities. In these gatherings, clan members can get a direct experience of being a part of a large and powerful whole, their clan. It is this feeling of being part of a larger, more powerful whole that gives visible and tangible things like totemic animals their power.

For Durkheim, the crux of social life was clan gatherings and rituals that kept the stories of the clan alive. Ahde-Deal collected many examples of family rituals that had a similar function. The stories that she heard were remarkably similar in Helsinki and Chicago, even though of course, these two cities are vastly different. One of the main differences was wealth. During the war and the postwar recovery period, Finland went through hard times that the United States never endured. For example, many Finnish families had donated their golden jewelry to the war effort in the 1940s and were wearing wedding rings made of cheap steel (high-quality steel was reserved for the military). For the Americans, this was unthinkable. When Ahde-Deal listened to these stories, she realized that material differences like these were spurious. Women in both countries saw themselves as the guardians of family treasures that connected them to their roots. The fact that she heard comparable stories in two separate locations made her more confident about her conclusion about the "powers" of jewelry and their social origins.

Of course, there is a world of difference between Durkheim's armchair studies of First Nations people in Australian deserts and modern, twenty-first-century women in Helsinki and Chicago. Yet the reasons why some things became important are conspicuously similar. Meaning lives in groups and collective experience. A story of how a grandmother coped with a miscarriage may help decades later, but it needs a vessel like a brooch to stay alive for generations.

Experience Becomes Decoupled from Technology

User experience became one of the primary concepts in design research around 2000, and the empathic group adopted it with ease. It gave the group philosophical and theoretical depth, as well as a connection to industry. In the early years of the concept, it posed a methodic challenge, however. It was not clear how to study it, the legacy from usability research was not enough, and psychological methods and frameworks (Hassenzahl 2004) required expertise that the group did not have. In response, the group started to explore methods for studying it. The search had started in the *Maypole* project, continued in the *eDesign* project, and culminated in *Empathic Design* (Koskinen et al. 2003), which explored methods from research literature—like experience prototyping, cultural probes, and storytelling—and formulated principles that the group has followed ever since. The book also showed that these methods worked in industry, which gave the group confidence in its footing.

Empathic methods led to an unintended consequence, though: they decoupled experience from technology. The group's precursors in the 1990s had studied the ways in which people interact with technology through user interfaces. In contrast, the interpretive approach to user experience pushed the group off from the use situation to study social processes that create conditions of use. Its methods were cultural probes, mood boards, ethnography, mobile platforms, and immersive methods that could capture experience, but it was not clear how they could capture structures beyond those situations in which experience takes place.

The group started to see evidence that its work was limited by its connection to a particular situation. It knew that situations come from somewhere; every seminar is unique, but university departments organize them regularly; a technician assembling a frequency transformer faces a unique situation every time, but his work brings him to installations daily, and he uses the same principles; every phone is unique after a few days of use, but the buying process in the shop is similar and standardized by every vendor. When these observations accumulated, new members of the group started to explore the domestication of design to understand some of the structural backgrounds of experience.

The decoupling of experience and technology bothered Battarbee, who wanted to find ways to couple experience and technology. The question that kept her awake was whether co-experience is different when technology mediates human interaction from a distance as opposed to when people are colocated. For instance, does being colocated somehow change the exchange over the phone? That is, if two people get a message, does their response differ from situations in which only one person gets the message? She could not find an answer to this question and came to agree with Bill Moggridge, one of the founders of the design firm IDEO. In his book *Designing Interactions* (2006), Moggridge describes the moment when he understood the importance of interaction design. He realized that when people use interactive technology, they focus on what they must do, *not* on the technology. If everything goes well, technology disappears into the background. Therefore, interaction design is important, he reasoned. Don Ihde's (2006) critique of the "designer fallacy" similarly resonated with the members of the group.

The decoupling of experience from technology was taken further in the domestication studies. For example, Ahde-Deal came to see the powers that she studied as beliefs, but she in no way suggested that adornments are magical. The secret of these powers lies in the beliefs and relationships that support them. Paavilainen learned that the meaning of design depends on how people define their home. When she shifted attention from objects to the social context, her observations started to make sense. The human side was clearly more important to these researchers than things and technologies. With these studies, the gap between technology and experience was growing from a ditch to a canyon.

But how would design look in an approach that decouples experience from technology? In the early years of the program, this question was prominent in Kurvinen's mind. His advice to companies he worked with was to create open-ended platforms in which people could create their own experiences and follow these platforms before rushing to decisions about product development. Do not dictate; give people freedom. Ahde-Deal (2013) raised the same question in her thesis. When she was sending her papers to conferences, her reviewers usually said that her research was charming but wanted to know why she kept doing design if her message was negative: what is the point of design if designers cannot create objects that will have the powers she was talking about? Her response was that she was urging designers to create designs that were open-ended enough to accommodate changes over decades. She also encouraged them not to consider themselves as professional failures if their objects were not loved by people, but rather to study and enjoy cases in which they were successful.

Empathic design came to acquire a clear social overtone in its early years. The group sided with an old item of design wisdom advocated by Mattelmäki. For her, designers can create conditions for experience, but they cannot design experiences. They can give cues to people with signs, symbols, sounds, lights, atmospheres, forms, and service sequences. People can and do use these cues, but experience is something that people fashion by themselves. The rest of the group found it easy to agree with her, and this answer was the same as the prototyping social paradigm by Kurvinen and colleagues (2008). It was a refreshing breeze of skepticism against the idea that designers should somehow script human life. The interpretive approach gave the empathic group a clear position on a main fault line of the design world: it prioritized human experience over technology. The implications of the ordering came back to shape the group in many ways over the next ten years, as the following three chapters will show.

Smart Products

Usability of Smart Products

Maypole

One-Dimensional
Usability

The Future of
Digital Imaging

Cardboard Mock-ups

eDesign

Design Games

Empathic Design

IKE

Luotain

Design Studio
in the Field

Mobile Image

Co-experience

Collaborative
Design

Radioline

Design Probes

Prototyping
Social Action

Women
and Jewelry

You are Important!

Dwelling with Design

Design for Hope

MorpHome

Electrone

Pasadena

IP08

Art of
Research

Prototyping
Interactions

SPICE Project

Design Research
through Practice

Isles and Plants

Against Method

Memories in Clay

Drifting by Intention

Folk Tradition

Ryijykutuja

What Happened

Novapro

From Disposable
to Sustainable

DWoC

Lost in Woods

3 Codesign and Commitment

This chapter shows how the group embraced codesign as a method of communication and commitment. By 2005, the group knew how to capture, describe, and translate user experience. These methods worked well, but they raised a communication issue. How can one communicate empathic knowledge to companies and government agencies and commit them to it?

The answer was codesign, a concept that was gaining ground internationally. Codesign was dialogic rather than researcher-driven, and it was better at creating commitment than user-centered design because it integrated users and other stakeholders into every stage of design research.

The group developed several approaches to codesign. One line centered around Danish work on design games and participatory design. The main outcome of that approach was the *Play Framework*. Another line explored the Brazilian roots of participatory design in the *Vila Rosario* project. A third line was collaborative. The *IKE* project sought a way to create a resident-centered approach for renovating apartment buildings. It led to a field station in the *Ave Mellunkylä!* project.

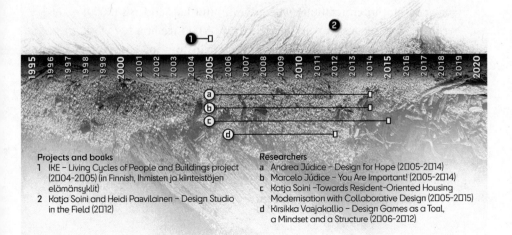

Projects and books
1 IKE – Living Cycles of People and Buildings project (2004–2005) (in Finnish, Ihmisten ja kiinteistöjen elämänsyklit)
2 Katja Soini and Heidi Paavilainen – Design Studio in the Field (2012)

Researchers
a Andrea Júdice – Design for Hope (2005–2014)
b Marcelo Júdice – You Are Important! (2005–2014)
c Katja Soini –Towards Resident-Oriented Housing Modernisation with Collaborative Design (2005–2015)
d Kirsikka Vaajakallio – Design Games as a Tool, a Mindset and a Structure (2006–2012)

Around 2003–2004, several members of the empathic group started to pay attention to warning signs between their interests and the interests of the clients of design. The signs came indirectly from design firms. On the research front, it was smooth going. Focus on user experience was serving the program well. It was abstract, but the frameworks and methods described in earlier chapters captured it well. Researchers from other disciplines found concepts like co-experience interesting and novel. But the situation was getting worse on the practical front. The group decoupled experience from technology and was developing concepts, methods, and processes to capture the former. But as an occupation, design works with particulars: it is about nuts and bolts, joints, bosses and ridges, line widths, placement of masses, and minimizing interference from motors. The gap was immanent and obvious. Cautiously but firmly, practitioners started to tell the group that it was hard to apply its research, and by 2005, these voices were getting louder. Theories and methods created in the early years of the program had solved one problem but led to another.

It was not that practitioners did not understand the value of the group's research. They knew it was important, and it had long-term benefits for the community. Still, the concern was legitimate. There was a need to define a research practice that would keep the theoretical and methodic advances but make them easier to digest in practice.

The solution was codesign. I am not aware of reliable histories of this approach, but I can give a sketch from the perspective of the group. It was clear that codesign was becoming important in European and North American design research, and the group saw it as a natural way to build design ideas with practitioners and manage communication with them. When the group was reading literature about this nascent notion, they quickly came across several interpretations. The group found Liz Sanders's (2000) early writings about codesign particularly useful. When reading her work, they were struck by how she was using simple participatory and contextual tools in her projects. Another reference came from California. In San Francisco and Palo Alto, IDEO was developing design thinking to keep customers committed to its prototypes, which were often shelved for years before customers put them into production (Brown 2008). The concept also appeared in England, where Stephen Scrivener, who taught at the Chelsea College of Art and Design London (now a part of the University of the Arts London), organized a research symposium in Coventry in 2000 (Scrivener, Ball, and Woodcock 2000). It was an engineering conference, but several papers from it found their way to the desks of the empathic group through colleagues who had taken part in the conference. In Scandinavia, codesign found proponents in Denmark and Sweden (Binder 2007), and it was also finding its way into Italian interaction design (Rizzo 2009). The group built on American references that were easier to apply than the European sources.

As this sketch suggests, it is hard to pinpoint any sole source or exact moment where codesign became a major concern in empathic design. Rather, it was a process that took several steps. As this chapter shows, its origins were in action research that the group had adapted to its needs at the end of the 1990s but that failed to gain a following in the wider design community. Its value was undeniable, however, so the group sought ways to keep its collaborative core but turn it into a design approach. The first to rephrase it was Tuuli Mattelmäki, who talked about design as a dialogue that creates understanding. Her work inspired Kirsikka Vaajakallio, a student of hers who explored design games to open design to nondesigners, and Andrea and Marcelo Júdice, who viewed her work through Paulo Freire's pedagogy and Pelle Ehn's participatory interpretation of Ludwig Wittgenstein's philosophy. These approaches found their most systematic expression in the hands of Katja Soini, whose path to codesign started in 2004 in the *IKE* project and culminated in her doctoral thesis (Soini 2015) .

Action Research and Its Problems

The idea of codesign is built on a decade-long history in Helsinki. Its origins lie in the notion of action research that was taught in the industrial design program at the end of the 1990s. Simo Säde decided that his research problem was introducing user-centered methods into a design consultancy, and he explored it in his *Cardboard Mock-Ups and Conversations* (2001) and built his approach on action research. His study became a precedent for the group and pushed action research into teaching. After joining the industrial design program in 1999, my first task was to teach action research to master's students. In the eyes of Professors Raimo Nikkanen and Juhani Salovaara, the leaders of the industrial design program, action research was a natural extension of ergonomics and usability, and Salovaara (1985) had studied the ergonomics of hairdressers from an action research perspective. There were few action-oriented research alternatives in design at that time, and they gave the task of teaching action research to me, the program's recruit.

In building the class, I started from design literature. I found a few disjointed references in design literature, but the crop was small. A handful of references emerged. Säde was exploring action research for the final study of his thesis, as did some others (Swann 2002). This body of literature built on higher education and social psychology rather than design, so I looked next at research in the social sciences, including consulting Kurt Lewin's (1946) classic texts to get the basics right. For Lewin, action research was a way to wean groups and communities from their counterproductive and harmful ways and replace them with better habits. Action research helped to understand the ills of the community, but its true power came from a change program built on research. The change program, Lewin taught, should build on social controls. Weaning young mothers away from smoking or using milk powder rather than breastfeeding by providing information and preaching seldom works. Changing the way that their friends think and talk about these issues does. If your friends tease you, scold you, and start to avoid you, you will consider quitting unhealthy behaviors— even if you do not believe what doctors tell you.

Action research was an attractive proposal that had been gaining some proponents in the local design community for years. Yet, there were reasons to doubt it. It was clear, for example, that the uptake of user-centered methods at the design consultancy that Säde was working with was limited. It was basically limited to one partner of the company, who saw it as a useful addition to his toolbox of approaches. My class was hardly a success either. Students followed my instructions because they had to, but many were dragging their feet, and some of the most talented design students outright rebelled. After a couple of years, I concluded that the distance between social psychology and design practice was just too long. The basic idea of action research was attractive and well tested, however. To change things, you must work with people and facilitate change rather than impose it from above. But it was also clear that action research did not work as a design tool. It felt too much like an alien implant. It was important to keep what was valuable about it, but there was also a need to define a better practice that was closer to student experience.

Codesign as Dialogue

A better practice was emerging from Mattelmäki's office. She wanted to find a concept that put designers and people on a more equal footing than in a usual design consultancy. The concept turned out to hinge on dialogue. For her, design research was a dialogue in which designers and clients jointly created interpretations in conversation. This view was different from the traditional interpretation of designers as professionals whose perspectives are more valuable than the layperson's. It also created a contrast to user-centered design, in which designers study subjects to learn about how they see the world, and then do the rest of the work in the studio. As her colleague Turkka Keinonen once wryly noted, in user-centered design, users are *used*. They may offer valuable insights, but designers have not been interested in them as full human beings: their value is instrumental. This was a model that Mattelmäki tried to leave behind. For her, dialogue did not rule out disagreement, but it did push her to view people as human beings rather than as subjects who had only instrumental value.

She formulated the concept in a series of papers about cultural probes (Mattelmäki and Keinonen 2001; Mattelmäki and Battarbee 2002; Mattelmäki 2003), which led to her doctoral thesis (Mattelmäki 2006). Her research was animated by the same question she explored in her doctoral study: do cultural probes work in product development even though their origins are in situationism, surrealism, and psychoanalysis? Her answer came from a series of industrial projects in areas ranging from finance to sport equipment manufacturing. Toward the end of her doctoral studies, she conducted a series of evaluation interviews to address this question. She concluded that the probes indeed work in industrial contexts, largely because they work as a dialogue between designers and participants. Her belief in dialogue was built into Keinonen's *Luotain* project, in which she was the lead researcher. It brought together thirteen companies to experiment with new methods.

Mattelmäki's concept also held the seeds to an even more radical redefinition of design. If design research is a dialogue between designers and clients, why not go all the way and extend this principle to other stakeholders as well? She suspected that restricting design to designers and clients was not enough. There are other stakeholders who must be given a voice too. She realized that there is no logical reason to restrict dialogue to only the early phase of design. On the contrary, designers should make sure that users and other stakeholders should be able to take part in all aspects of design, including research, concept design, sketching, and even prototyping. The spirit of dialogue opened a logical path that pushed the empathic group step by step toward codesign—or, in plain English, designing together.

The most important source of information during Mattelmäki's transition to codesign was probably Liz Sanders, who had started to talk about codesign a few years earlier (Sanders 2000). Her writings were evocative and exciting to Mattelmäki, providing her with a way to manage meanings that arise during the design process. Sanders's theory was too abstract for the group, which did not share her knowledge of psychotherapy, but her methods were another story: the group could easily assimilate them. In a few years, the group was well positioned in the evolving field of codesign. When the journal *CoDesign* was established in 2004, its very first article was written by Battarbee and me (Battarbee and Koskinen 2004), and Mattelmäki has contributed several times to it.

Design Games as Codesign: The *Play Framework*

The idea of dialogical design provided a starting point to the industrial designer Kirsikka Vaajakallio, who became Mattelmäki's doctoral student. She was intrigued by the latter's ideas but wanted to find a structured approach to dialogue. In her doctoral thesis *Design Games as a Tool, a Mindset and a Structure* (Vaajakallio 2012), she spiced up the group's legacy with a Danish influence. She repeatedly visited Copenhagen's venerable Danish Design School to learn how Professors Thomas Binder and Eva Brandt used design games in codesign. Their interpretation became the foundation of their famous (but now defunct) master's program on codesign. In Denmark, she also became familiar with Jacob Buur's group in Sønderborg, Jutland, which had direct connections to Scandinavian participatory design.

The Helsinki group had earlier experience with design games. Games and gamelike elements were explored in several studies in the 1990s. The group was aware of Brenda Laurel's *Computers as Theater* (1991), and Keinonen had explored collaborations with an opera director while working at the Nokia Research Center, and he knew the pros and cons of working with artists. A good example of this early work was the puzzle interview, in which users are asked to solve puzzles or unusual problems to measure characteristics like creativity, flexibility, and intuition (Keinonen and Soosalu 1997). The group was also aware of Katja Battarbee's (2004) research into fun and play, as well as the computer scientist Giulio Jacucci's doctoral thesis (2004), which built on Victor Turner's anthropology of theater. A bit later, in Aalto's Media Lab, Katriina Heljakka's doctoral thesis *Principles of*

Adult Play(fullness) in Contemporary Toy Cultures (2013) explored play and toys. It was also possible to find many design studies featuring prototypes as games. When you create a laboratorylike environment of the cockpit of a forest tractor (Säde 2001) and bring in experienced operators to study user interfaces, you are in fact creating a make-believe world that lasts for only an hour. During this time, the rules of the outside world are invalid; the play world of the test and its rules define reality for the participants.

Vaajakallio's interest in games made the group aware of its playlike practices by placing design games on a philosophical and sociological foundation. She built on the work of the Dutch historian Johan Huizinga and the French sociologist Roger Caillois, whose classic work on games and play had clarified how play transports the participants to another world and how these temporary belief systems are supported (Huizinga 1950; Caillois 1961). She was also reading Richard Schechner's (2006) performance studies, which helped her to understand more extended performances, and Ludwig Wittgenstein's (1953) "language games," which gave her philosophical foundations to study the gamelike aspects of design research. Finally, Erving Goffman's (1961) sociological analysis of a "situated activity system" provided her a way to study transitions into Huizinga's "another world" in serious contexts like surgery and rituals (see also Ylirisku 2013). These transitions were also familiar to her from Battarbee and Koskinen (2004) and many practical design pieces in the community. Several colleagues, for example, had been designing textiles and ritual vessels for the church, jewelry and emblems for families and professional associations, and stages for theaters. Her theoretical foundation led her to think that the common thread behind these transitions was make-believe, and they could be analyzed as instances of games.

She started her studies with Make Tools (Sanders 2000). These were (mostly) Velcro blocks that could be combined like Lego blocks to support interviews about future devices. For Liz Sanders, they helped people to avoid words when revealing their unconscious needs. After becoming interested in the theory behind games and play in Denmark, her attention shifted from situations into the structures that produce them. Learning from Binder and Brandt, as well as from Jacob Buur, who simulated product development scenarios with Dacapo, an experimental theater based in Odense, Denmark, she began to see codesign in terms of games that could simulate institutions (Buur and Larsen 2010).

Vaajakallio developed her approach to codesign in projects with three companies. The first was the multinational elevator manufacturer Kone, for which she designed a *Character Game*, which was intended to help it internalize perspectives from various users. For OP Bank, she designed two games: the *Storytelling Game* facilitated service concept development, and the *Project Planning Game* helped to set up a joint direction to the bank's product development. She replicated these two games in a third project with a strategic service design consultancy. For each game, she planned a flow from explaining the basic idea to getting interaction moving, activating all the participants, and creating a "magic circle," in which participants convene to do a task by following printed rules that detach them from their regular world. She captured these games by keeping the materials, making notes, taking photographs, and shooting video.

For her, games became venues of codesign. To construe codesign events, she found tools from Liz Sanders, Bill Buxton, Tony Salvador, Giulio Jacucci, Jane Fulton Suri, and Simo Säde (Sanders 2000; Buxton 2007, 267; Salvador and Howells 1998; Säde, Nieminen, and Riihiaho 1998; Iacucci, Kuutti, and Ranta 2000; Ehn 1988). A particularly valuable part of her games was the role that these tools came to assume in her make-believe worlds: she did not see her props and process charts as experimental stimuli, as they had often been seen in the research literature. The object of her interest was human interaction, which was enabled by the tools. Theater and games gave her a rich vocabulary to distance herself from the idea that the tools were stimuli. Participants became performers and actors rather than subjects, and they had roles to play. Designers became producers, directors, and choreographers. Supporting staff became technicians and gaffers. Other participants became spectators, fans, congregations, juries, or spectators. Just as in theater, the magic happened onstage, between the actors and the audience, and it was enabled by a larger and often invisible organization.

The *Play Framework*, as she came to call her approach to codesign, was a toolbox of methods, but it was also a mindset and structure for her. The framework gave her a way to plan and describe her design games. From her Danish colleagues, she learned to prepare the design games carefully, which was in marked contrast to the group's open-ended philosophy of research design (Kurvinen, Battarbee, and Koskinen 2008). Most codesign events in the empathic program had previously been prepared as workshops rather than as theater, which was a much more structured metaphor that also suggested scripting and dramatizing the events.

Vaajakallio's study had many implications for the group. The *Play Framework* gave the empathic group a better understanding of Binder's and Brandt's versions of participatory design. It also gave the group a structured approach to designing codesign sessions (Mattelmäki, Brandt, and Vaajakallio 2011; Vaajakallio and Mattelmäki 2014). Her theoretical explorations helped the group to find theoretical links to intellectual luminaries like Wittgenstein and Huizinga, and her methodic explorations helped the group to incorporate imaginative artistic techniques into design. The step from her interpretation of design games as make-believe to scriptwriting and design fiction was short.

Vila Rosario: Paolo Freire and Empathic Design

Another interpretation of Mattelmäki's dialogical design emerged from Andrea and Marcelo Júdice, two Brazilian designers who joined the group in 2004. Rather than gamifying design, they had a distinctly Brazilian take on her ideas. Through their work, the group started to see some of the limits of its approach. Codesign offered the empathic group a powerful method for integrating stakeholders into design, but it quickly became clear that it had limits. It took place in Helsinki, a well-off capital of a Nordic welfare state, and it mainly collaborated with well-funded and sophisticated information technology companies. These facts introduced many northern European assumptions into empathic design. For example, most product developers whom the group worked with had a background in computer science or engineering. The companies they worked with were well off. They had resources for experimenting with methods like cultural probes and fieldwork, and they knew how to subcontract research. And the users were people who were educated and in good health, and they had time to play. The group mostly worked in its hometown and did not have to deal with the complications of cultural differences and languages.

Many of the limits became visible when the Júdices joined the group in 2004. They wanted to use their design skills for social good (A. Júdice 2014; M. Júdice 2014). Their work initially focused on how information technology could be used to improve public health in Brazil. During the first year of their doctoral studies, they decided to focus on Vila Rosario, a small village about fifteen miles north of Rio de Janeiro. Specifically, they focused on the detection and prevention of tuberculosis, which was exceptionally prevalent in the village. Importantly for our story, they faced several limits of empathic methods in the village. How would cultural probes work if half of the population of the village were illiterate and only a fortunate few could afford to access computers and internet connections? How can information technology be designed when it was well known that the public computers previously installed in the village had been stolen after a couple of days? Who could be trusted if locals say that local gangs and drug kingpins are more honorable than the police?

It was clear to the Júdices that research in Vila Rosario had to be based on a solid understanding of the village and its inhabitants. The methods had to be contextual and build on firsthand experience of the village. The researchers soon realized that methods like probes would work in the village only if they were adapted to local conditions and administered through local organizations. Many patients could not read, but nuns and nurses could.

The conclusion was clear enough. The key to success lay in knowing the local context. The Júdices had to get this understanding in order to make their methods work. They came from the better part of the town, and as they had no local knowledge of the village, they could only trust their process. They needed a support structure to improve their chances of success. The Júdices ruled out some candidate partners, including the local government because the villagers did not trust it. The church fared better in this respect, but again, it had little interest in research that did not support its religious mission. The Brazilian League against Tuberculosis (*Fundação Ataulfo da Paiva*, FAP) had a program in the village, however, and this proved to supply the right context. It worked with two city-based doctors who had been working pro bono in the village for years and knew its problems intimately. The villagers trusted the program, and its foot soldiers were health agents recruited from the village.

The Júdices were surprised to see that their most imaginative methods worked particularly well in the village. For example, Giulio Jacucci's "magic things"—wooden or foam blocks that could be used to prompt stories that elicited requirements for mobile devices—were easy to use and playful enough (Iacucci, Kuutti, and Ranta 2000). The only people who were wondering what they were about were the doctors. Trained in the life sciences, their reaction to the magic things was bafflement at first, but this turned quickly into appreciation when they saw how they worked. The same applied to the probes: the villagers appreciated the effort put into them. Health agents administered them, and they worked well. From Mattelmäki, the Júdices learned that it is important to check the interpretations that came from probe returns, so they conducted ethnographic research in the village. For this, they developed more surreal devices to elicit stories in fieldwork. For example, when they asked people to tell them how a "Good Fairy" could solve their problems, they heard that the villagers trust doctors from *telenovelas* more than real doctors. In response, they created a story world that mixed features from soap operas and reality to guide their design work.

The design program the Júdices built consisted of eleven designs and an identity for the clinic they worked with. Most designs were simple featuring posters about the importance of hygiene, comics about the symptoms of tuberculosis, and protective clothing for health agents to make them identifiable on the street. The job of these agents was to walk around the village in the hot tropical sun to diagnose cases of tuberculosis and support treatments. The Júdices' research results told them to direct their efforts toward low-tech design and to push high-tech solutions into the background. They discussed and sketched ideas with the villagers and health agents to codesign the details together. The purpose was to make sure that the villagers felt that the creative output of their work felt right and was their own (figure 3.1).

They also did extensive tests in the community before producing the designs. The outcome was positive. Their designs were not sophisticated by the standards of the traditional design world and would never win awards, but they worked well in the local campaign against tuberculosis. They also led to an approach that could be replicated.

Their next target was Vila Mimosa, an impoverished area west of downtown Rio de Janeiro, near Maracanã, Rio's storied soccer stadium. When the sun was shining and its car repair shops, eateries, and other small businesses were open, the neighborhood was poor but charming. When the night fell, though, thousands of prostitutes and patrons roamed its streets, bars, and brothels. Diseases like tuberculosis were rampant, and the area was also a significant source of HIV infection in the surrounding metropolis. The Júdices worked there with doctors and the prostitutes' union to disseminate information about health, nutrition, and education. They also replicated parts of

Figure 3.1

The *Vila Rosario* project. (1) Probes and two returns; (2–3) probe returns; (4) ethnography; (5) a codesign session with the villagers; (6–7) testing potential designs; (8–9) two designs out of the eleven that were used; (10) three characters created for design. (Picture credit: Andrea and Marcelo Júdice, 2004–2014.)

their approach in Windhoek, Namibia. After their fieldwork was over, they have continued to help and guide health agents at Vila Rosario through social media ever since.

In true empathic manner, the Júdices did not stop here, however. They went on to explicate two frameworks to describe their research approach. Following the program's open-ended philosophy, the frameworks were not defined in the beginning of the study. Instead, the plan was to build on the best research practice, conduct research and design in Vila Rosario, and then find out how the studies could fit into the program. After finishing fieldwork in the village, the Júdices sought a way to connect empathic references to their Brazilian roots. Surprisingly, the crucial theoretical link came from Sweden. One of the best-known participatory designers, Pelle Ehn, had had a Marxist past as a student. In his doctoral thesis (Ehn 1988), he described two classic participatory projects from the 1970s and placed them in a philosophical context. In an intriguing footnote in the beginning of his thesis, Ehn mentioned that the Brazilian teacher Paolo Freire (2005) had inspired participatory designers in their early years (Ehn 1988, 8–9). The question that Ehn's thesis tried to answer was why his simple participatory methods—like cardboard computers—had been so successful years earlier. One of the answers he gave was that they helped to create a mutually understandable language game between graphic workers and his colleagues. This term came from the philosopher Ludwig Wittgenstein (1953). I knew a good deal of the literature that Ehn was referring to, including Wittgenstein's philosophy, and I was also familiar with Freire's pedagogy from my student days at the University of Helsinki. When I saw a reference to Freire and Wittgenstein in Ehn's thesis, I thought they might provide a basis for the two theses.

Freire gave the Júdices two principles. The first was easy to adopt: design materials must build on local sources and be intelligible in the context of their use. The second was more complicated: it was important to make sure that research helps the poor but does not glorify them. Freire's pedagogy insisted that liberation is not real if it raises anyone, including the poor, into a position of power. In Hegelian and Marxist terms, turning former slaves into masters would not amount to true liberation. This notion became the basis of Andrea Júdice's *Design for Hope* (2014). Marcelo Júdice based his *You Are Important!* (2014) on Wittgenstein instead. The difference in approach reflected their roles in the process: she was the main researcher, and he did most of the design work. In his thesis, he used local characters, themes, and symbols to create a joint language game with the villagers, which he saw as being a crucial tool in taking part in the village's "form of life," in Wittgensteinian language. The outcome was a piece of research in which design was clearly local, in contrast to much of Brazilian academic design. As the Brazilian journalist and scholar Adélia Borges (2011) has noted, academic Brazilian design has German roots. Its ascetic technological functionalism is incongruous with indigenous Brazilian design and its way of life, materials, sounds, smells, culture, and sense of what is good.

This observation led to a number of ethical questions. Like practically all approaches to design research, empathic design is a First World creation. In a town like Helsinki, research faces few limitations. Having a typhoid or yellow fever outbreak in Helsinki would be world news. Also, several generations of wealth have eradicated many dangerous household practices. For example, almost everyone uses separate knives to cut fish, poultry, and vegetables, which reduces the risk of spreading foodborne diseases. Even more prosaically, everyone can read. Designers can give written instructions and create user interfaces using words.

The *Vila Rosario* project showed that the validity of these assumptions depends on local circumstances. Vila Rosario was a poor community in which access to clean water was a luxury, yellow fever a reality, poultry often contaminated with salmonella, and rubbish was dumped illegally close to water sources. The illiteracy rate was high, and electricity was routinely stolen.

Who deserves design? Is design a prerogative of people living in Amsterdam and West Hollywood, or would it be better to rethink its audience— perhaps by starting to design for those who go to bed hungry rather than for those who find a semiotic analysis of the *Tree Trunk Bench* on Staalstraat an intellectual exercise worth a couple of pages in a zine? As Anthony Dunne and Fiona Raby argued so well in *Design Noir* (2001), a good deal of design follows the Hollywood blockbuster model: it tries to please everyone, so it pleases no one. Yet there is a paradox in their work: its audience is even more elite than the audience of ordinary design. After all, Hollywood blockbusters and scissors by Fiskars are distributed globally and reach the poor, while critical design does not. What the Júdices did was different. They met with people who had hardly heard of design, and they did their research to serve these people, not to generate debate in centers of global wealth. Theirs was critical design research from the school of Paolo Freire, but with a Scandinavian middle-of-the-road touch.

Commitment and Collaborative Design

Codesign had become the empathic group's dominant process by 2010; it still built on its user-centered legacy, but its research process had become collaborative. Codesign had replaced the authority of the designer with a more equal process, and the group had begun to see itself as a facilitator of change rather than its driver. Its methodology became more democratic as the idea that designers are experts was pushed to the background. Codesign had also embedded the design process much better into its context, be it a government agency, a medical setting, or a global corporation. Unlike action research, it also felt right to designers. The dialogue at its heart gave empathic designers powerful tools to manage stakeholders. Still, it was not clear why it worked. Projects like the *Play Framework* and *Vila Rosario* helped to create a better understanding of the possibilities and the limits of codesign but could not explain why it worked so well.

An answer to this question came from Katja Soini's doctoral work, which followed *IKE*, a project already mentioned in the opening section of this chapter. *IKE* had formulated a process for a resident-centered renovation of Helsinki's suburbs (Virtanen 2005; Soini 2015). Many of these suburbs were built in the 1950s and the 1960s, and by the turn of the century, some parts of them were at the end of their life cycles. Some high-rises were demolished, but most were refurbished to remove dangerous materials, improve insulation, and replace water and sewage pipes, as well as electrical risers. Typically, this work was done by builders who paid little attention to inhabitants and their needs: on the contrary, this industry was notorious for its indifference to its customers. Knowing the reputation for excellence in user-centered design of the University of Art and Design, Helsinki, the Ministry of Environment gave the contract to it rather than urban planners or civil engineers.

The result was the *Living Cycles of People and Buildings* project (known as *IKE*; 2004–2005). "IKE" is an abbreviation of "Ihmisten ja kiinteistöjen elämänsyklit," the original Finnish name of the project. The project, funded by the Ministry of Environment, brought together progressive engineering firms and the government to find ways to turn major renovations of apartment buildings resident-centered. Soini ran a series of seventeen workshops that brought together various stakeholders to find better ways to renovate buildings. In cold climates like Finland, this process had tradition-ally been invasive and very costly. *IKE* sought ways to give residents a voice in the process. It proved that workshops were a useful way to bring indus-try and the government together to innovate new processes. The group had always worked collaboratively with other designers and clients, but the novelty of *IKE* was that it opened up the design process to a much wider group of stakeholders.

Soini wanted to connect inhabitants and industry to create a new, resident-centered renovation process. To do this, she had to find a way to create a community of stakeholders whose worlds were disparate and interests often at odds. Her method for creating such a community was running a series of workshops (Soini and Pirinen 2005). In the footsteps of user-centered design, she collected user data to use as props in the workshops. She invited inhabit-ants and their organizations, city and government officials, and builders to the workshops. She described the user data, structured it into a series of ques-tions, and asked her participants to find workable solutions to the problem. The outcome of the project was a booklet that described a new type of pro-cess that would give inhabitants a larger role (Virtanen 2005). It was turned into formal guidance by the Ministry. Over the next two years, the project led to changes in national statutes about renovation. One government minister adopted several ideas from the report, and it led to fifty-one government-funded projects and experiments around renovations. It had a major impact, and Soini wanted to know why.

An obvious answer was that that government and the democratic pro-cess behind it had power. When the parliament changes laws, industry must react to avoid sanctions. However, this answer says little about what role of design research played in changing the government.

The first ingredients of an explanation were theoretical. The sociologist and management scholar Rosabeth Moss Kanter had a theoretical insight that supplied a starting point. In *Commitment and Community* (Kanter 1972), she had asked why some religious communities endure for centuries. In answer-ing her question, she described how these communities manage to drag people into the community and exclude interactions with outsiders whose beliefs are different. Exclusion creates commitment to the ideals of the com-munity. Following Kanter, Soini's working hypothesis was straightforward.

Her workshops had brought together several significant stakeholders (the government, construction industry, engineering firms, nongovernmental organizations, inhabitants, and universities) to create empathy between these groups. She believed that this would help to create a vision that everyone could accept, regardless of the divergent standpoints that a society might have. This shared vision would then create commitment in ways that ordinary policy processes could not do. Methods were secondary; the crucial thing was striking a balance between the goals of the government, the industry's multibillion-dollar interests, and the needs of someone like an average empty-nester named Hannah, who lived in her apartment after her husband, Leo, had passed away. As Soini came to see it, industry and the government assumed that they knew what Hannah needed. Their method was consulting standards and building regulations that were often old and detached from the reality that she faced in her everyday life. In contrast, Soini believed that every major renovation project should be driven by an empathic understanding of people like Hannah.

This was the hypothesis she wanted to test in her doctoral work. Soon after *IKE*, she had children, and so she published her thesis in 2015, ten years after the project had finished. This gap gave her a unique opportunity to track its aftermath. She studied the network around *IKE* repeatedly to map its impact. The network had over seventy participants, she learned. Some of them were well-connected lobbyists who had never been seen in the workshops, nor were they mentioned in preparatory documents. By tracking the history and the impact of the project, she learned that her study had been an important element of a process that had been in the works for years before she was commissioned to work on it, and the process continued several years after it was finished. It was the Ministry and an industry network that kept the legacy of *IKE* alive. Several engineering firms that took part were thought leaders, and the chief executive officer of a large construction consultancy became a champion of codesign in the construction industry. Through her interviews, she saw how the industry slowly started to change its ways and how the implications of the project spread throughout the country in just a few years.

She also realized that although Kanter's theory had opened her eyes to commitment, it was only a Blumerian sensitizing device. The theory was able to explain how religious communities work in rural settings in the United States, but her theory did not work well in the open, competitive environment of contemporary industry. Although it was not perfect, it still led Soini to what can be called an institutional theory of design impact. She concluded that her work was a success because of the network and political connections behind it, and designers should understand the institutional embeddedness of her work to improve its chances of success. As a young designer, she had not been aware of this embeddedness, even though it was the main reason for her success.

Although she came to doubt whether Kanter's theory of commitment could explain her case, she also saw features that the theory could explain. For example, she learned that a significant part of commitment is giving up social contacts to people who were not members of the community. In Soini's case, the giving-up process was decidedly modern. As you give up theories, perspectives, industrial, and organizational practices, you make decisions that are financially and technically calculable. Giving up is a rational process that builds on a belief that something better is achievable. As she came to see it, commitment is a lengthy process in which existing practices are redefined step by step.

She called her approach "collaborative design," and it was different from Mattelmäki's dialogical approach to codesign. Soini saw empathy as a gradual process, in which a professional designer needs to dive into a network and lose her distance before returning to a professional level. After she has internalized the feelings of the stakeholders, these feelings can inform research, but distance is also necessary, she believed. She studied psychotherapy to learn about how she could manage the tension between getting close while keeping distance, and to learn about techniques for viewing her own feelings in a larger social and processual context. For Mattelmäki, empathy was a way to create dialogue. For Soini, it was a dialectic of getting close while maintaining distance, and an ability to see emotions in a broader political context. If the impact of a project depends on the network around it, it is crucial to understand this network and keep it in mind during the design process.

Studio in Suburbia: *Ave Mellunkylä!*

Another assumption at work in codesign was scrutinized in the early 2010s. Codesign was still a designer-led exercise when it occurred through games and workshops. These were planned by designers, who created a structure of meaning that lay at the bottom of their design activities. What if codesign activities were taken out of the studio to places that designers could not control?

This question was explored by Katja Soini and Heidi Paavilainen in the *Ave Mellunkylä!* project in 2012. Mellunkylä, a suburb in East Helsinki, is a leafy area consisting of several old villages that grew into suburban centers after World War II. It is linked to Central Helsinki by subway and freeway, and it has become ethnically diverse over the last thirty years. Large parts of it consist of privately owned homes, but it also has housing projects known for social problems, and there are occasional clashes between ethnic groups. At this time, about 25 percent of the population was foreign born, and its unemployment rate was 15 percent. In some schools, only 10 to 20 percent of pupils spoke either or both of the two official languages of the country, Finnish and Swedish. For people in Helsinki, Mellunkylä epitomized poverty and social problems, but the actual facts often contradicted this impression.

In the early 2010s, the City of Helsinki started a program to invigorate its suburbs. Most of these suburbs are found about 8 miles from the center of town. One of the projects was granted to Soini and Paavilainen (2013), who set out to study the neighborhood for several months. With a group of students, they aimed to create plans for reinvigorating the neighborhood, which was heading into an era of renovation. Most of its apartment buildings were built in the 1960s and the early 1970s. While still in relatively good condition, they would need extensive repairs and renovations in ten to twenty years. However, the inhabitants lacked the political resources of the better parts of town. The project was based on Soini's collaborative design and enriched it with Paavilainen's trend analysis. The latter was trained as a textile and fashion designer and was familiar with future studies and trend research.

The duo rented an apartment in Kontula, a notorious subcenter of Mellunkylä, as well as an empty shop in the area's aging shopping mall. They turned the shop into a four-room studio that also functioned as a community space. Upon arrival in Kontula, they gave interviews to local newspapers, distributed 500 handouts describing their aims, built a web portal, and participated in resident events. The team stayed in the neighborhood for eight weeks. During that time, they created media contacts, arranged visits by politicians, and organized a meeting between the locals and the chair of the Parliament, the second most powerful official in the country.

Thirteen students took part in the project. They invented several codesign methods to gather data about the neighborhood, including a *Graffiti*

Wall in one of the rooms of the studio, and a *Tree of Tolerance* that the inhabitants could use to express what they loved about the neighborhood. The team organized public events in the studio to gather materials and codesign futures with the locals. The outcome was a series of four scenarios describing the future of well-being in Mellunkylä and exploring the impact of multiculturalism, tourism, ecology, and climate change.

The main benefit of being in the neighborhood was that the research team faced many of the difficulties in the same way the locals did. In no time, they had a stockpile of stories to share about bad internet connections, unreliable postal service, and adverse weather patterns that caused delays in public transport. As the team became familiar with the neighborhood, the local inhabitants came to know them and saw that they were not the average City Hall researchers who come from better neighborhoods for brief, one-time visits. The relationship between the team and the inhabitants became empathic: each side developed a sense of the other. Having a temporary home in Kontula gave the team a better understanding of the neighborhood and ensured that their scenarios were rooted in its reality. For example, the researchers learned that the locals had a clear sense of aesthetics, but it was practical, situated, and less abstract than in university classrooms. It had to be personally meaningful. Mellunkylä was not a place for sleek modernism that catches the eye of design juries.

When reflecting their experience, Soini and Paavilainen explored the implications of having a studio in the field. The noted that when on site, they could not limit their creativity to any single activity; many things happened simultaneously. They could not control their schedule; they had to react when people had problems. They could not rely solely on a process; a lot of improvisation was required—they had to leave behind the safe world of a university class or contract research. In the field, the team needed to face people and organizations they could not foresee, and it had to interact with people in whom academic discourse has seldom been interested. Rigorous research methods and model-making techniques were fine, but social skills were more important.

That fact of having been there also meant that people in Kontula saw them differently from other consultants who had tried to revive the neighborhood. These projects had typically been one-time projects by researchers and designers in search of slum romantics. Locals saw their home, quite rightly, as a leafy and peaceful neighborhood that was good for raising

children and retiring—except for a few public housing blocks that had undeniable social problems and an occasional group of rowdy teens, alcoholics,
and drug addicts in the mall. In fact, one message that the research team
kept hearing was the same one that Marcelo Júdice had heard years earlier in
Vila Rosario: the simple fact that students from a famous design school were
willing to leave their homes and live in Kontula gave inhabitants a boost
in self-confidence. The experience also led to collaboration between the
neighborhood association and the neighborhood association of the upscale
city center neighborhood of Töölö, as well as to an institutional partnership
with Aalto University. The project gave the researchers "an understanding of
the suburb and people's systems of meanings, and it made them rethink its
future" (Soini and Paavilainen 2013, 96).

Cooptation and Commitment

One of the persistent problems of design has been uptake among its customers. This problem also exists in design research. You can do marvelous
research, draft clever presentations, and whip up enthusiasm and excitement, but as soon they are on their own, people in customer organizations
maintain their routines and do next to nothing that researchers suggest to
them. Practicing designers have dealt with this problem in many ways over
the years. So have design researchers, who had to find ways to communicate their results better so they would be palatable to their audiences.

As we have seen in this chapter, the predecessors of the empathic group
built on Kurt Lewin's (1946) action research. It tells designers to change
the social setting to change things rather than rely on the hypothesis that
a good piece of design will somehow automatically lead to change. Action
research was one of the springboards of empathic design, but it was discarded by the group, mainly because the thinking behind it came from the
social sciences, and it was not exciting to students. Its undeniable merit,
however, was that it shifted the focus of the empathic group from a studio-
based approach to a social context.

For the empathic group, a better answer was codesign, a notion that
was emerging in the early years of the century. The group realized that by
bringing audiences to the design process, the ownership of ideas became
shared, and there was no need to convince the audience about their value.
The path from dialogical design to codesign took a few steps, though. The

first one was taken by Mattelmäki (Mattelmäki and Battarbee 2002; Mattel-mäki 2006), who spoke for a dialogical approach in which designers, users, and other potential stakeholders work together under a project framework created by designers. Her approach was at odds with two beliefs: design-ers are experts, and their expertise justifies a top-down approach to design research. Her belief in dialogue paved the way to an interpretive under-standing of design, but it was designer-led. Designers were given the right to frame the situations and events in which dialogue was to happen.

In her footsteps, Katja Soini's *IKE* project showed how a workshop-led design process can shape policy, Kirsikka Vaajakallio (2012) explored design games to enrich codesign processes, Andrea and Marcelo Júdice (2014) inte-grated codesign and fieldwork in Brazil, and Soini (2015) developed her notion of collaborative design by establishing a field station to codesign communities in East Helsinki with Heidi Paavilainen (Soini and Paavilainen 2013). By the early 2010s, the group had built several codesign frameworks, all grounded in Mattelmäki's work and the empathic methods that had been explored before that.

In retrospect, the most important consequence of the dialogical approach that she advocated was perhaps that when the international design research community started to talk about codesign, the empathic group did not have difficulties onboarding the movement. The notion was familiar, and it felt at once right.

Still, it was hard to understand what codesign means beyond the obvious answer, which is that it means "designing together." As the French design scholar Annie Gentes once quipped to me over a drink in Paris, one prob-lem with the Latin languages is that you can add the prefix "co-" to almost any verb to create a new term. As her sarcasm implied, creating innova-tive terms is easy; giving them a precise meaning is not. For the empathic group, codesign was an opportunity, but it also needed a way to define it for its own purposes. It did not try to write a definition but started from the spirit of the term. For the group, codesign created an obligation to open design to parties who had never been part of it: children, cancer survivors, radio listeners, politicians, neighborhood associations, and so on. Also, the notion shifted attention from the first stages of design to a broader, partici-patory model. User-centered design had taught the group to study people to inform concept creation and prototyping. This changed with the notion that it would be illogical to keep nondesigners away from concept creation, sketching, prototyping, or evaluation—an observation captured later by the title of the Italian design scholar Ezio Manzini's (2015) book *Design, When Everybody Designs*. Codesign opened design research and turned it into a joint exploration from the very beginning to the very end.

As codesign varied, it also became confusing (Mattelmäki and Sleeswijk Visser 2011). Soini's collaborative approach provided one way to clear up these confusions. She was attracted to codesign through a personal doubt about whether design is needed at all, or whether it is only part of the industrial complex that had created many of the world's current problems. Codesign offered a way out by helping her to ground her design ideas into the needs of a community. She wanted to facilitate change rather than just design together. The collaborative ideal that she advocated offered a new role for her: her job was to create conditions in which people themselves can formulate their problems and seek solutions to them. Her approach defined the relationship of design researchers to their audiences and had clear affinities to what Sanders had been doing in Ohio.

One thing that was driving the empathic group toward codesign was the changing nature of design. The group worked in projects that few design firms would have seen as billable. Only Vaajakallio worked with smart products and services. The Júdices worked with a nongovernmental organization, and Soini with a joint venture of the public sector and industrial opinion leaders. Some master's projects using codesign techniques had customers like the Finnish Broadcasting Company. The push toward collaborative design came from a need to work with issues that needed solutions on a scale that is difficult to manage using the traditional tools and processes of industrial design. There were invariably multiple independent stakeholders that had many, often incongruent perspectives and interests, especially when the public sector was involved. Often these stakeholders had significantly more power than design researchers. These projects introduced the group to the "garbage can" decision-making processes (Cohen, March, and Olsen 1972) that frustrated them. One problem with garbage can processes was that parties can jump in and out of them at will. How can one manage a purposeful process when commitment varies and can be incredibly low?

The solution coming from Soini's work was cooptation. This term came from a study about the decision making of the Tennessee Valley Authority (TVA) by the classic organizational scholar Philip Selznick (1949). With cooptation, Selznick referred to the TVA's habit of putting people from opposing parties into decision-making positions to manage opposition and preserve stability. By coopting opposition, it was able to manage free riding by creating commitment among its stakeholders.

This sums up what Soini's collaborative design tried to achieve. It was to be a consensus-building process that brought many interests together to build commitment and trust. In *IKE*, Soini was dealing with a billion-dollar challenge that not even the government could manage alone. Rather, the challenge required the collaboration of a whole range of organizations from banks and engineering firms to lawmakers. Cooptation was a practical way to bring together an array of organizations, small and large, private and public. One benefit of this framework was that it also created connections to contemporary sociological and economic theories and to political philosophy—all fields that have dealt with consensus and conflict for decades, if not centuries. These references did not remove the problems, but they did make her approach intelligible to public administrators, lawyers, and politicians at the helm of government.

Smart Products

Usability of Smart Products

Maypole

One-Dimensional
Usability

Cardboard Mock-ups

The Future of
Digital Imaging

eDesign

Design Games

Empathic Design

IKE

Design Studio
in the Field

Mobile Image

Luotain

Collaborative
Design

Radioline

Co-experience

Design Probes

You are Important

Prototyping
Social Action

Women
and Jewelry

Design for Hope

Dwelling with Design

Pasadena

Marahome

Electrome

IP08

Art of
Research

Prototyping
Interactions

SPICE Project

Bikes and Plants

Design Research
through Practice

Memories in Clay

Against Method

Folk Tradition

Drifting by Intention

Runkohirua

What Happened

Novapro

From Disposables
to Sustainables

DWoC

Lost in Woods

4 Interpretation and Radical Innovation

This chapter describes how the group developed methods to support calls for radical innovation.

The interpretive approach gave the empathic group tools to capture the world as it is in order to improve the user experience. After 2005–2006, however, calls to support innovation started to grow louder internationally.

The group responded to these calls by connecting to craft designers, artists, and the ArtCenter College of Design in Pasadena, California. These connections inspired the group to adopt techniques from design practice and art. They also led to an empathic mode of analysis that shifted it from data to the emotions that it elicited in researchers.

The key project inspired by these collaborations was called *SPICE*. It created concepts for subway stations with scriptwriters, scenographers, and urban sociologists. Imaginative methods were also explored in *Opening the Electrome*, which hacked electric meters in apartment buildings in Karnataka, India; and *Bicycles and Plants*, which introduced fiction to ethnography.

The final section of this chapter studies how interpretive aesthetics can respond to the call for radical innovation.

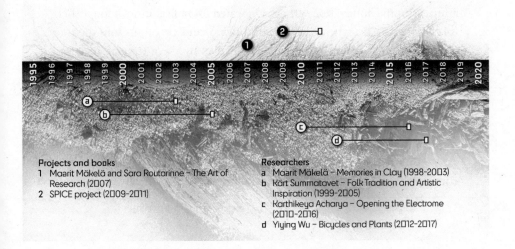

Projects and books
1 Maarit Mäkelä and Sara Routarinne – The Art of Research (2007)
2 SPICE project (2009–2011)

Researchers
a Maarit Mäkelä – Memories in Clay (1998–2003)
b Kärt Summatavet – Folk Tradition and Artistic Inspiration (1999–2005)
c Karthikeya Acharya – Opening the Electrome (2010–2016)
d Yiying Wu – Bicycles and Plants (2012–2017)

After a few years in the making, Roberto Verganti published *Design-Driven Innovation* in 2009. His book presented a meaning-based theory of design innovation. According to him, designers create innovation by changing the meanings of things rather than by inventing technology. His theory was built mostly on famous Italian examples from the 1980s, including the Memphis group's furniture, and Alessandro Mendini's and Stefano Giovannoni's designs, which helped Alessi to transform kitchens into playgrounds with a toylike design language.

It was a sign of the times. Verganti's tour de force was the way in which he linked design to meaning. It summarized well the ways in which a good deal of leading Italian design had been developing since the birth of radical design in the 1960s. Its forte was marked by an uncanny ability to see familiar things with fresh eyes. The weak points of Verganti's argument were equally obvious. His data reflected the industrial context of Lombardy of the 1980s, in which companies like Alessi and Driade and movements like Memphis designed familiar products like chairs and tables by shifting their meaning (e.g., by designing kitchenware as if they were toys). It pushed designers to develop design innovations from new ways of thinking rather than from science, technology, materials, or business ideas. It was also used to belittle hard-won knowledge in user-centered design.

His way of reducing design to semiotics was novel, but another part of his argument had been articulated in business literature that the empathic group had been reading years earlier. The leading business authors from California and the East Coast of the United States had been arguing for years for the need for drastic, radical innovation rather than approaching innovation in an incremental fashion. Some proposed that business ought to focus on creating disruptive innovation, others advocated for radical innovation, others told business to become agile and fail forward, and yet others told them to change the rules of the game rather than play by them (Dyer, Gregersen, and Christensen 2011; Hamel 2002; Hamel and Prahalad 1994; Kelley 2001; Maxwell 2007; Prahalad and Krishnan 2008). The empathic group followed this literature through its Irish colleague Peter McGrory, who specialized in strategic design, and I was familiar with the sociological literature that had inspired these business authors (Järvinen and Koskinen 2001).

Another trend that was shaping international design research was noncommercial, but equally important. Traditional art and design schools had been toying with the idea of turning design practice directly into research. Critical design was gaining traction in England, and designers in California had started to talk about design fiction (Dunne and Raby 2001; Sterling 2005). Another common term was "practice-based research," which had roots in England, which suggested that design research should build on making things, crafts, and perhaps artworks (Scrivener 2000, 2004; Biggs 2002). During the first decade of the twenty-first century, it developed into the Art of Research conference (Mäkelä and Routarinne 2007), which alternated between London and Helsinki, and craft research, which had a journal dedicated to it in 2010. This research was not commercial, but it was making the call to bring more creativity to design research louder.

The problem that these developments posed to the empathic group was that the very heart of its approach was at odds with them. The new calls wanted to ground design in imagining what could be (Simon 1969; Cross 2001), but the group's interpretive approach seemingly anchored design to what exists rather than imagination (e.g., wishes, fiction, stories, dreams, and even nightmares). Since its inception, however, the empathic group had been building creative tools that would help it to face these calls without sacrificing its theoretical legacy. Ten years later, it was obvious that by focusing on imagination, the group's interpretive approach had become more versatile. This shift took several years and built on international sources, but the immediate starting point was the wealth of resources with roots deep in Helsinki's design legacy. The jolt came first from Pasadena through Tuuli Mattelmäki and me, and later through new doctoral students, especially Karthikeya Acharya and Yiying Wu. The idea of building on design-based methods, however, was a natural continuation of the culture of the design school in which these researchers were working. These methods represented a move away from the descriptive methods of user-centered design.

In the Shadow of Design Legacy

The group was working in a famous design school that had produced so many design classics that it was hard to keep track of them. Most professors they worked with were award-winning designers, and many of them had designs featured in museum collections. When former colleagues included people like Marimekko, Alvar and Aino Aalto, Timo Sarpaneva, Eero Aarnio, and more recently the Showcrash group, artistic and design references were easy to find, readily accepted, and encouraged. Almost every student was trained in plastic arts and drawing by well-known artists. There was no route away from the ethos. Working against it in the way that Simo Säde (2001) had done—he printed his doctoral thesis simply, in Microsoft Word, rather than giving it to a skilled graphic designer—became just another artistic statement. It was impossible not to be an artist in an art school.

When the empathic group was forming, many of its members had studied craft disciplines including ceramics, glass, jewelry, and fashion. For example, Esko Kurvinen became an accomplished researcher, but his true passion was glass blowing. These members of the group had grown up next to a culture of unique pieces, one-offs, small batches, experimental prototypes, and signed objects (Lovell 2009). Many industrial designers in the group were also working in crafts and had a plenty of experience with fictional methods from the student days. For example, in the *CODE* project (Ahonen 1996), students had explored product semantics and the use of collages, and their task was to create physical models of toasters, blenders, and other kitchen appliances. They had to choose a form language to do this. The outcome was a series of models through metaphors like toys, power tools, porn, and sports.

Some students had also taken part in *The Workshop* (1995) and *Keittiössä* (2003) projects with Centro Studi Alessi, the upscale Italian kitchenware manufacturer. In workshops like these, students learned to use industrial forms to make statements. An example of a student project was a spoon to stop drug addicts from stealing spoons from coffee shops: it had a syringe-shaped hole in the bottom, which make cooking drug doses impossible. Another project taught filmmaking skills. It created concepts for a smart phone enabled with global positioning system (GPS) technology for Benefon, a now-defunct phone manufacturer. Benefon wanted to display the possibilities of its novel technology in the world's largest consumer electronics fair at the time, CeBIT, in Hannover, Germany. (CeBIT comes from Centrum für Büroautomation, Informationstechnologie, und Telekommunikation, or in English, "Center for Office Automation, Information Technology and Telecommunication.") The students produced four 90-second films with film students, which were scripted and acted and included shots from firefighters inside burning buildings and skydivers over the wintry plains of Lapland. Another precedent was *Designing with Video,* a book that

Salu Ylirisku published with Jacob Buur. It was user-centered rather than fictional, but it relied heavily on creative and playful design techniques (Ylirisku and Buur 2007).

The group also had members whose research was more in debt to artistic imagination than observation. The ceramic designer Maarit Mäkelä was particularly important. Her doctoral thesis, *Memories in Clay* (2003), explored feministic and phenomenological concepts through three exhibitions of ceramic artworks. Her thesis showed the group that art, craft, and research can blend seamlessly. These exhibitions explored female experience through a method that she called "the retrospective gaze." The first step of her method was an artistic process that defined a theme that led to an art exhibition. The second step was a retrospective analysis of the exhibition and the preceding process. This step identified the theme for the next pair of an exhibition and the subsequent retrospective analysis. The outcome was a series of three experimental exhibitions. Her first exhibition built on a critical analysis of media, the second on an auto-ethnography of her own body, while the third elaborated an autoethnography of four generations of women in her family through "autofiction." Her study showed that art and research can indeed meet in a methodologically sound way, as a physicist sitting in the audience whispered over drinks to me.

From this background, it was easy enough to develop research methods to explore the imaginative side of design. The creative mentality drove Mattelmäki's (2006) probe studies, but it was also clear in the visual scenarios, films, and many prototypes that her students were doing. Similarly, it found its way into research through methods like the "Good Fairies" of Andrea and Marcelo Júdice, who also built a cartoon-style twin of Vila Rosario to create designs that felt right in the village (Júdice, Júdice, and Koskinen 2015). Even quirky methods found acceptance in the group and its design environment if they served the purpose. An ability to repurpose existing forms was not in short supply in an old art school.

Super Studio: Learning from Pasadena

The decisive jolt for breaking away from the grip of user-centered design can be attributed to a collaboration with Pasadena's revered ArtCenter College of Design. I visited it in 2006–2007 and Mattelmäki did so in 2007. Its

Media Design Program had been running a yearlong research class called Super Studio, a brainchild of Brenda Laurel, but in 2006–2007, it was taught by Lisa Nugent and Sean Donahue, two faculty members with studios in Los Angeles. It was run like a design studio. Every week, students brought their research and designs into the classroom, where they were critiqued. The class had a research exhibition at the end of the first term and a design exhibition at the end of the second term. Students were mostly young industry veterans from the West Coast, and their final thesis was usually an exhibition that told the story of their research (Nugent et al. 2007).

For Mattelmäki and me, the most interesting thing about the studio was that although it was a research class, it built on design imagination rather than methods from social science textbooks. Some of these methods had been published in Laurel's *Design Research* (2003). In 2006–2007, the theme of the studio was "Biospheric Voices": how Los Angeles residents experience nature across various ecologies in the city. The 2006 studio was based on cultural probes, but these were more imaginative than the disposable cameras, postcards, and writing tasks that had become a canon of probing in the first half of the decade. The students went beyond the canon, however. One student, for instance, wanted to know how people hear nature in Los Angeles. He covered the windows of a few homes with transparent film and gave markers to participants. Then he asked them to write what kinds of natural sounds they hear. These sounds ranged from babies, dogs, and wind to coyotes in the foothills. The biosphere indeed was talking to the Angelinos—but in different ways, depending on whether they lived in a house in the foothills or in a flat downtown.

The most important single piece of inspiration for the empathic group was Miya Osaki's thesis, *Retellings* (2008). Osaki explored the Japanese American experience during World War II and its implications for Japanese identity in the United States. She communicated her research with stories of war and about the trauma of internment camps in an exhibition. Her main visual elements were old photographs that were blown up to human size, and evocative texts from her family members. Her diverse inspirations ranged from Pieter Breughel's paintings and Yoko Ono's *Cleaning Piece* to the science fiction author Bruce Sterling and discussions with faculty. When Mattelmäki saw the exhibition, she wanted to bring the idea back to Helsinki (figure 4.1).

Figure 4.1
From Miya Osaki's *Retellings* (2008), at ArtCenter College of Design, Pasadena. This project was an important inspiration for the empathic group. people.artcenter.edu /~osaki/retellings/. (Picture credit: Miya Osaki.)

Through the Media Design Program, Mattelmäki and I were exposed to the emerging idea of design fiction a few years before it started to gain traction in design research and human-computer interaction (HCI). It became popular in design-oriented HCI a few years later, but it had a longer history. Appropriately, Los Angeles, the capital of storytelling, played a key role in this buildup. Fiction as such had been a standard element in design for decades (Kymäläinen 2015). For example, Disney had helped the National Aeronautics and Space Administration (NASA) to create a vision of space travel in the 1950s, the Milan-based Memphis movement that created a revolution in design in the 1980s that was largely built on fiction (Radice 1985), the philosophy of the studio was also rooted in Brenda Laurel's (1991) vision of computers as theater, and computer scientists had been using scenarios as design tools for years (Carroll 2000). However, design fiction was formulated for the first time in *Shaping Things* (2005), a short book by Bruce Sterling. It evolved during his residency in the Media Design Program in 2005. The belief that fiction is important was strong in the program because of Sterling, and because of Peter Lunenfeld, an accomplished author, one of the design world's great verbal acrobats, and a core faculty member of the program (he is now at the University of California, Los Angeles).

Design fiction also struck a familiar chord in the group through Mika Pantzar, a former professor in industrial design in Helsinki, who had written a book about how fiction had been driving future homes (Pantzar 2000). His book illustrated some of the ways in which artistic visions of the home had shaped the market and how companies like Disney have defined consumer expectations on the silver screen. The book, written in Finnish, argued that fictive ideas can drive design. Through his work on mobile multimedia and connections to the School of Cinematic Arts and the Annenberg School of Communication of the University of Southern California, I also met Julian Bleecker, who was working on what became the most widely read pamphlet about design fiction (Bleecker 2009). Southern California was the cauldron of design fiction, and the Media Design Program supplied a rich window into it. When the center of design fiction moved to England a few years later, the empathic group had no difficulties related to it.

"What If": Spiritualizing Space

After decades of delays and lawsuits, Helsinki was expanding its subway line in the second half of the first decade of this century. Helsinki's subway is generic, designed with maintenance and efficiency in mind, and its sensory and cultural environments were poor. Its surfaces are easy to clean, materials standardized, and access to stations uniform, as are the look and feel of the stations. Its brutal lack of character caught the attention of Mattelmäki, the design leader of a project called *SPICE—Spiritualizing Space* (2009–2011). It became the first project in which her research agenda was clearly shaped by her experience in Pasadena. Its goal was to supply an alternative way of thinking about subway stations in Helsinki by exploring wayfinding, customer journeys, service blueprints, and customer touch points through storytelling and role-plays (Mattelmäki and Vaajakallio n.d; Viña and Mattelmäki 2010; Kankainen et al. 2012).

The main reason that she wanted to rethink the subway was that the extension was also changing its nature. As it was extending from Helsinki to the neighboring Espoo, its demographic base morphed significantly. Helsinki is a densely built European town with significant social differences. Espoo is low-rise and suburban. It is also well off. While Helsinki suffers from many metropolitan problems like homelessness and poverty, Espoo is the richest large municipality in the country. When passengers catch the subway in East Helsinki, they leave the densely built, relatively poor, and politically left and populist-leaning suburbs of East Helsinki. They go under the city center, the country's commercial, administrative, and political hub. When they surface in Espoo, they see another world entirely. It is leafy, well off, car-friendly, and politically green and conservative. The journey through these worlds is a small-town version of François Maspéro's trip through Paris in *Les passagers du Roissy-Express* (1990) or Jacques Jouet's voyage in *Poèmes de metro* (2000).

Mattelmäki was struck by the obvious mismatch between these different life-worlds and the functional design philosophy of the subway. The surfaces were flat and made of materials that are easy to clean, caverns for the stations were concrete, and signage consistent, as were the colors. The layout of the stations depended on whether they are underground, above the surface, or in a shopping mall, but beyond these three ecologies, truly little locality was built into them. Riding the subway was like descending into an underground city that imposed one-size-fits-all values on commuters. The contrast to other subway systems familiar to Mattelmäki was intriguing. What if the stations in Helsinki could reflect their environment as the Hollywood/Vine station in Los Angeles does, with thousands of film reels telling you that you are in Tinseltown? Or Arts et Métieres in Paris, which is designed to tell passengers that they are about to enter the story of scientific and technological progress? What if we could capture the spirit of a place and use that as a base for design in Helsinki in an analogous manner?

It was this question that Mattelmäki wanted to explore. She set out to capture the spirit of the proposed new stations by exploring and probing neighborhoods around them. The *SPICE* project was largely built on well-tried techniques like probing and codesign workshops. Importantly, however, she also wanted to communicate her results in a format richer than paper and slides.

The main concept was "spirit." It was meant to create some distance from the argument that the environment of the station should somehow be directly mapped into the design of the station. *SPICE* wanted to avoid using direct references like the film reels in the ceiling of Hollywood/Vine. The aim was not to tell stories of the locals either, such as by romanticizing the former working-class neighborhoods in the east, or by using stories of business conquests in the office jungle of Keilaniemi in the west. The concept of spirit was a way to maintain a distance from overly straightforward design solutions. It forced her to study planned station areas to capture their multiple meanings.

SPICE was a collaboration with computer scientists, sociologists, scenographers, and scriptwriters. In the hands of the filmmakers, design scenarios did not start from realistic stories of how people enter the subway, but from drama: What if your tie gets stuck in the escalator when you are rushing to a job interview? What is the worst-case scenario? What if you meet the love of your life while trying to rush to a job interview? These kinds of reality twists had the benefit of not being tightly bound to existing things and experiences. Scenarios like these turned everyday reality into explorations into existential dilemmas, surreal experiences, intoxicated escapades, and nightmares. As Mattelmäki has noted, these what-if scenarios could turn an empathic understanding of everyday life into fictional futures and imaginative design concepts (Mattelmäki, Routarinne, and Ylirisku 2011).

The most visible outcomes of the project were exhibitions that communicated the concepts (Mattelmäki, Brandt, and Vaajakallio 2011). Mattelmäki's team built exhibitions to trigger empathic responses to its design ideas. The team created open-to-interpretation posters and booklets that invited viewers to step into the shoes of other people. These communication experiments aimed to make viewers curious and amazed about the wealth of human experience behind routine and mundane realities (Fulton Suri 2003). They also aspired to trigger conversations that could create meanings that developers and designers could share. Over the next few years, Mattelmäki used these techniques in several service design projects that she became involved in through the newly formed Service Factory funding scheme at Aalto University. Her approach to services was distinctly imaginative, in contrast to the prevailing user-centered and systemic approaches that were popular in engineering, computer science, and design communities in Europe.

Inspiration as Analysis

The empathic group was working in a design school. There were no problems in getting its new imaginative methods accepted at the school, but
analysis remained a problem. How can materials be analyzed without
destroying the creative, evocative nature of the methods that produced
them? Can there be a creative path from observations to design concepts
that is still acceptable in the world of research, which requires clarity?

The answer built on local design practice that flipped the question
around. Master's students once asked their professor, Juhani Salovaara,
what he was looking for when he was grading their designs. "Intuition,"
he answered. The students were baffled and left wondering how they could
know whether they had intuition or not. What they did not understand
was that he was talking about *his* intuition, not theirs. How could we bring
Salovaara's way of thinking into research?

The problem was not the school, but the wider research world. The group
knew plenty of methods for collecting data, but it was not clear to them
how to do it. Statistical methods were out of the question except for simple
percentages, pivot tables, histograms, and pie charts, but design students
often got percentages wrong and had never heard of binomial distributions,
logit models, or odds ratios. Almost by default, the group had used affinity walls as a method of analysis (Beyer and Holtzblatt 1998). The method
was straightforward. Researchers wrote their observations on sticky notes
and then started to cluster them. These notes were easy enough to group
and regroup on the walls of the studio until they started to create patterns.
Affinity walls were attractive for many reasons. They were intuitive, familiar
to industry, and well established in design firms. Furthermore, they avoided
the complicated terminology and processes of sociological alternatives, like
grounded theory, that had inspired them (Glaser and Strauss 1967). The
walls were practical, but they were also becoming boring because they did
not feed design imagination: they were still grounded in the social sciences.

A design-specific way of seeing analysis was forthcoming. Many leaders
in design research had started to explore ways to use imaginative design
methods in analysis. An early sign came from a talk that Bill Gaver gave in
Helsinki in the winter of 2002 (Gaver 2002). He had been invited to talk
about cultural probes to a seminar organized by the *Luotain* project, which
introduced probes and contextual design to thirteen companies. Gaver

described his method of analysis as "gossip." He put his probe returns on a table, browsed through them, and then started to imagine what kind of person had sent them. After putting together a picture, he could start to design for that person—regardless of whether the picture he had created was right or wrong. Designers like Mattelmäki could easily see that the rationale behind the word "gossip" was the key, like Salovaara's intuition. But how can we explain "gossip" to students who needed methods because they lacked experience?

Here, the Pasadena experience came in handy again. In the Media Design Program, I had seen several examples of how students could use their visual skills in analysis. For example, in 2006, Yee Chan and Serra Semi gave several families an empty globe and asked them to cut and paste text and images onto it from the Sunday edition of the *Los Angeles Times*. After receiving the globes, they photographed them and started to collage them. First, they used simple categories like "teenagers" and "grown-ups." Later in their analysis, they collaged data from several participants and categorized them in two hemispheres—the Global North, talking about industry, pollution, luxury, and travel; and the Global South, talking about climate catastrophe, poverty, and disease. They printed these hemispheres on a magnetic base, painted the wall of the studio with magnetic paint to create a space for an exhibit, and covered the magnetic wall with white paint the next day. When the doors of the exhibition opened, visitors were given magnetic speech and thought bubbles and markers, and they were asked to reorganize the collages and to add their own thoughts to them. People who came to see the midterm show became coanalysts. The criteria of validity shifted from truth to evocativeness and from researchers to visitors.

This experience gave me enough reasons to rethink analysis in design. I wanted to find a way to think that would make designers feel at home with this concept. As I saw it, the students were not analyzing structures and processes in their probe returns in the same way that a science student would have done. Instead, they were interested in following how it affected them. They took the data seriously, but they did not try to minimize their impact on analysis, as a scientist would have done in the name of minimizing method variance. Instead, they treated their own hunches and feelings as indications of what might be interesting. They discussed their cases and clustered them as any scientist would, but then they decided what was interesting by reflecting on what *they* found important, fascinating, or intriguing.

When their thoughts and feelings started to converge, they knew that they had found a theme that was robust enough for design. "Intuition" again.

As the brilliant design anthropologist Ken Anderson once said to me, he calls this form of analysis "user-inspired research." Anderson smirked when he said the "user-inspired." At first, I did too. The idea of analyzing data by following the feelings that it elicits in you went against everything that I had learned as a student. I had learned to think that analysis is a game in which data must have a fair chance to prove that researchers are wrong. Soon, however, I realized he had seen something similar many times before in design schools. It was how Tuuli Mattelmäki treated her probes, it was how experienced designers like Kärt Summatavet (2005) understood analysis, and when Turkka Keinonen returned from industry in 2002, he called his first design class "User-Inspired Design."

The best analogy I could find was countertransference (Greenson 1967). In Freudian psychoanalysis, the analyst listens to the patient's free associations. Over the course of the therapy, the patient starts to project emotions to the therapist, who then can start to trace emotional patterns from them. This is known as "transference," a main tool of psychoanalysis. By observing these emotional patterns, the analyst can start to piece together an image of what has shaped the patient's mind. In classical psychoanalysis, these patterns are usually traced to childhood—the authority of the father, the mother's love (or lack of it), sibling rivalry, and so on. These patterns may have been useful in childhood but often become dysfunctional in adulthood. The analyst's job is to help the patient to build a more realistic self. A main danger—a scientist would say "contaminant"—is "countertransference." Inevitably, the analyst starts to develop emotions toward the patient. These emotions could cloud the analyst's ability to be neutral and must be controlled, or the therapist may reinforce the patient's dysfunctional emotional patterns instead of helping the patient. To stay objective, psychoanalysts themselves must go to therapy.

As I saw it, countertransference was a useful analogy for understanding how several members of the empathic group analyzed their materials. Of course, there is no transference in empathic design: bits of data on an affinity wall or a video cannot project emotions toward researchers. Countertransference, on the other hand, does exist: researchers do project their own emotions onto people and events in data. By observing these emotional impulses, they can see what tickles them in order to find ideas that might serve as potential launchpads for design. Psychoanalysts control countertransference by talking to other analysts; design researchers control it by talking to their peers. Some ideas fall flat in these conversations; others get everyone excited. Emotion—ranging from excitement to awkwardness—shows where minds should go. Experience and other designers supply a reality check.

Was this answer good enough? Certainly not, if the standard for analysis is set by textbook-level science. The method that my colleagues were using, however, was familiar to me from another context. Every good researcher I knew worked this way in the first stages of analysis. They had talked about hypotheses rather than inspirations, of course, but imagination was a quality that any researcher must have. Besides, I thought, designers will test their ideas by building concepts and prototypes and testing them. From my student years, I also knew of a paper by Donald T. Campbell (1975), who had done more than perhaps anyone to turn psychometrics into science. Campbell had started his career in the 1950s by criticizing Freudian psychoanalysis for its lack of rigor: if you analyze only one case, it cannot prove you wrong. He revised his view years later, when he realized that the true sources of rigor in psychoanalysis are its theory, stock of case histories, and dialogue with other analysts. These provide a ground against which observations are accepted or discarded. In brief, rigor is not based on those mathematical constraints that Campbell had believed in as a young man. The source of rigor was the community that made sense of ambiguous observations. If this reasoning was good enough for a leading psychometrician, it should be good enough for design researchers, I eventually concluded. Since then, I do not smirk anymore when I hear people talking about "inspiration," and I hope Anderson could accept my reasoning.

Opening the Electrome: Hacking Meters in India

The University of Art and Design had seen several sustainable design initiatives since the 1960s through the contacts of its former president, Yrjö Sotamaa, with Victor Papanek. Sustainability resurfaced in the school at the end of the 1990s and the 2000s (see, for example, Siikamäki 2006; Kohtala 2016; Marttila 2018). Sustainability became a theme in empathic design at a doctoral level when Karthikeya Acharya joined the group. Behind his interest was a familiar conjecture from India. What if the 1.3 billion inhabitants of India one day consume as much energy per capita as people in the West? Early signs are alarming: as the standard of living in India has risen, the new, emerging middle class increasingly wants to live a Western, energy-intensive lifestyle. Whether there are ways to make people aware of their unsustainable lifestyle choices was Acharya's question in his thesis *Opening the Electrome* (2016).

Trained in architecture and interaction design, Acharya was a talented ethnographer. He did fieldwork in four locations in the southern Indian state of Karnataka. He realized quickly that calculating how many watts people consume and building a smart-phone application to remind people to cut their energy consumption had been done many times, to little avail. To begin with, few people knew how to interpret scientific measures. Another problem was that even if they did, they needed a reliable benchmark that they could understand. Also, there were already measures that most people relied on—most notably, electricity and gas bills that supplied citywide averages for comparison. His problem was real, but not solvable by feeding traditional measurements to the phone or to a web site. As his supervisor, I kept repeating an old political wisdom that suggested employing taxes, the price mechanism or social controls like laws to change society rather than place our trust in the power of information or knowledge.

Acharya's attempt to create sustainable consumption patterns shifted to informal social controls. His approach was initially inspired by Anthony Dunne and Fiona Raby's *Placebo* project (Dunne and Raby 2001), Dunne's *Hertzian Tales* (2005), and Johan Redström's *Switch!* project. Redström's prototype artwork *Erratic Radio* intrigued Acharya (Ernevi, Palm, and Redström 2005). It measured electromagnetic radiation in its environment. If the radio sensed that there were many appliances running around it, it started to jump between stations and emit electrical noise. To get back to the program, the user had to switch off some of the appliances. His main theoretical reference was the architect Kenneth Frampton's (1983) essay about critical regionalism, which told him to start design from local objects, cultural patterns, and social forms. He took Redström's and Frampton's ideas and applied them to the context of what he had learned of empathic design. His approach became an empathic elaboration of these two authors on how to design for deenergization of India.

How this mélange worked in practice was defined by fieldwork. Acharya started by studying mobile technologies as awareness platforms in the cities of Manipal, Mangalore, and Bengaluru, all in Karnataka, and continued to study festivities in these locations. His first interest was defining how to measure the material footprint of middle-class lifestyles. The focus was on the middle classes because they are the source of most emissions and their lifestyle speaks best to the aspirations of the large, poor majority. It soon dawned on him that fieldwork would not be enough because few consumers usually grasp in detail how they use electricity at home and how their use compares to their neighbors' use. To enrich his fieldwork, he created a series of ad hoc and bespoke design interventions and deployed them in the field.

Acharya's most relevant intervention was a series of meter hacks. He got permission to hack electric meters in several middle-class homes and apartment buildings. He also got permission to publish the results in social media in order to make people in the buildings aware of their own consumption compared to their neighbors. How would this knowledge affect energy consumption, and how would it help address his large-scale research question about energy usage in India? Would people discuss their consumption, and would they start to find ways to share energy-saving tricks? Would social controls and shame kick in and lead to reduced consumption? Acharya's fieldwork concentrated on finding answers to these questions.

What were the main results of the meter hack? He saw that publishing energy consumption over the social networks was important. It gave his participants an opportunity to get a meaningful hyperlocal benchmark for their consumption. He reasoned that a citywide or neighborhoodwide average of electricity bills was too abstract to be meaningful and assumed that if people could see their consumption and compare it to other households in the same building, it would be much easier to see whether their consumption was roughly normal. This turned out to be the case. The figures led to stairway conversations in the building. For example, some participants wanted to know why their consumption was higher than their neighbors by collating things they knew about their neighbors. These facts were things like the number of children, appliances at home, and personal history. Learning that a family used to live in the Arab Emirates led to the assumption that their home was filled with appliances and they ran air conditioning around the clock. Acharya's designs turned energy consumption into a matter of local sense-making.

Acharya also brought some of his observations back to the studio in Helsinki, where he created several installations to prompt discussion about energy consumption. The most important installation was *Light Is History*, which was a rough-looking energy booth installed in Hakaniemi Square in Helsinki. The purpose of the installation was to create discussions that would help him to discover yet unnoticed aspects of the India experience. He continued exhibiting his work all through his thesis project, in part inspired by Mattelmäki's *SPICE* project.

Through his approach, Acharya's writing became perhaps the first critical doctoral thesis in the empathic group. Since the release of *Material Beliefs* (Beaver, Kerridge, and Pennington 2009), a project in London that explored the possible implications of emerging biomedical and cybernetic technologies with imaginative prototypes, exhibitions, and debates designed to move scientific research out of the laboratory into the public space, critical designers had stressed the importance of debate as a tool of social change. They wanted to create designs that disturb people in order to raise debate about design and its underlying assumptions. For them, design was a vehicle to create debate, which then would lead people to realize that they need to change their ways. Closer to Helsinki, this idea had also been voiced in the Iaspis Forum in Stockholm (Ericson et al. 2009). Acharya's research took place in the context of fieldwork in India, rather than in the galleries and museums of in cities like London and Stockholm. His audiences were not museumgoers or well-heeled design academics, but his methods were conversational at their core.

Empathic Fiction: *Bicycles and Plants*

Conversation was the starting point of another critical and fictional thesis arising from the empathic program, Wu Yiying's *Bicycles and Plants* (2017). She studied the critic Grant Kester's ideas about conversational art. In two books, he had documented how art had turned away from its traditional media and performances to conversation and collaboration (Kester 2004, 2011). Wu was also inspired by Acharya, my observations about relational art and conceptual aesthetics (Koskinen et al. 2011; Koskinen 2016), Ezio Manzini's (2015) social innovation agenda, and ethnomethodology that she learned from her supervisor, Jack Whalen. She wanted to question the assumptions behind service design which, as she saw it, was too reliant on the market. It assumed that services were designed to satisfy consumers, who were willing to pay for them. Her underlying interest was the social underbelly of services: trust, friendliness, and altruism rather than sharp-elbowed market capitalism.

Her pilot study was a bike repair shop in central Helsinki, in an old rail-yard, now demolished. The location is excellent: today, the area has a music hall, a new city library, several company headquarters, and some of the country's most expensive flats. The parliament and the National Museum can be reached on foot in three minutes. Back then, however, the yard was wasteland, and the shop anything but upscale. Its service was free. Repair technicians worked nonprofit and served you only if they liked you. If you entered demanding service, they threw you out. The same happened to visitors who were obviously well off. On the other hand, if you were nice and they could see that you were a poor student, you went to the front of the line. Access to services was a matter of negotiation rather than money and hinged on the usual access hierarchy for services. This is not to say that the service had no problems. It had obvious sexist biases, for instance. Wu wanted to study the shop because she saw it as a natural breaching experiment (Garfinkel 1967; Crabtree 2004). It was her window into studying how access to services can be a function of negotiation and social comparison rather than money, status, and wealth (Wu, Whalen, and Koskinen 2015).

Wu's main research intervention was a series of five Plant Hotels. She set up four real-life hotels in 2014–2015. Three were in Helsinki—one was in a nonprofit neighborhood gallery, one at a university, and one in a senior citizens' center. One was located at a design conference in Stockholm. Her fifth hotel was fictional and took place in the future. It explored whether plants could ease the tension between North and South Korea. The plant hotel was a simple concept: you can bring your plants to a "hotel" and give instructions about how they should be taken care of. Nobody was serving you. You had no guarantees that someone would take care of your plants. There were no fees, and the hotels were not organized by any legal body. These were experiments in volunteering and altruism. Would people trust their beloved plants to a service concept built around altruism and goodwill? Can people adopt the perspectives of people they do not know and take care of their plants?

The outcomes from her four real-life experiments were robust. The hotels worked. People brought plants to the hotels and left them there when they were traveling. Passers-by, professors, conference participants, and retirees did take care of them. A total of 63 people brought 153 plants to her hotels, and she saw at least 105 people taking care of them during the four months that it took to run the facilities. A few plants died, but this did not cause a stir. The service was mostly successful, and it created bonds between almost 200 people who had never met each other, and usually did not meet during the study either. The participants created theories of people they did not meet, and these imaginations were enough to glue the community together. Services like hotels could build on trust rather than on the formal legal controls behind a transactional service model.

These four hotels had a clear empathic fingerprint. Wu's study was clearly interpretive, although it also built on Garfinkel's (1967) ethnomethodology rather than on symbolic interactionism. It started with a short pilot that made her question sharper. It went to the study proper, which consisted of her plant hotels, which she followed with field research methods. The study was clearly empirical, design-based, lightweight, and not without a healthy dose of humor. Her way of analyzing what she saw was certainly interpretive. Yet her study also built on design fiction that was gaining ground in human-computer interaction (HCI) while she was in the beginning of her research (Blythe et al. 2002; Lindley and Coulton 2016).

Her last hotel was pure fiction, however. It explored a frontier that empathic design had not explored before. It was a speculative study of how a plant hotel could function on the militarized border between North and South Korea. Her script, written in 2016, began with this sentence: "Plant Hotel 5 opens at Panmunjom at the border of North and South Korea from May 2020" (Wu 2017, 143). She took the floor plan of the famous visitor center on the border and redrew it as a plant hotel. She developed a version of the hotel, security protocols, travel sites that talked about it, and fictional social media dialogues about events in the hotel. She assumed that political tension along the border would ease in 2018, and a part of the script included the political environment. She also created media images of the hotel and wrote a fictional newspaper article about it, including how it did its small part to ease tensions between the two countries.

Wu explicitly noted that since Plant Hotels 1–4 were real, she saw the border hotel as a secondary case. What did this fictional case add to her study? In a broad sense, it was a fictional study of empathy and its consequences. The story was set in a realistic frame: a hotel like this would mostly be discussed in social media and on travel sites. Her insistence of the importance of the political environment added another layer to the story. The hotel was to be a subplot in the tension between the Koreas and the larger powers behind the two governments. Behind all these subplots still lay an assumption about conviviality mediated by plants. The gist of her hotel was the relationship between people from both sides of the border and the way in which her plant hotel would shape their understanding of each other. It was a story of how people could be attracted to imagining the faces of the enemy, and how the process of imagination could ease the conflict. It was a conversational design piece in which the conversation was fictive (Kester 2004). The idea was that there would be mutual role-taking, even though the conversation would happen in two bubbles that would never intersect. Yet it is possible to argue that this imagined conversation could help to manage the tension between the two countries (see Wu and Koskinen 2021).

The Korean hotel was fictional, but it was doing what she intended by raising questions about imaginary assumptions that kept the services going. Why were her hotels working? Wu's interpretation was that ultimately, conviviality drove this hotel just as much as it did for her other four hotels. She may have been on the right track. People involved in running the hotels did not live together and seldom knew each other. There was no dialogue; any dialogue was only imaginary. Still, it was clear that people trusted their plants to unknown people, and plant caretakers responded positively to this act of faith. It was this web of assumptions that kept the hotels going, even though there were only the sparsest rules: no money was exchanged, no promises were made, and nobody could be held accountable if a plant died. Neither were there any real sanctions. There were no advertisements, and nobody was greeting participants with a fake smile at the entrance.

The power of her study as a whole—and Plant Hotel 5 in particular—lies in its interpretive implications. They helped her to study some of the imaginary social ties that keep services running. Her study gave her an opportunity to be critical of the idea that companies, governments, or other formal organizations must organize services. Behind any service is a more primordial but invisible human layer. Her *Gedankenexperiment* ("thought experiment") helped to make this primordial layer observable.

The Aesthetics of Interpretation

As this chapter has argued, the interpretive foundation served the empathic group well from 1999 to 2007 by giving it concepts and methods for describing experiences and introducing the results of its studies to industry.

The design landscape had been shifting. Importantly, the business literature had started to talk about blue-sky thinking, creative destruction, failing
fast forward, and disruptive innovation (Dyer, Gregersen, and Christensen
2011; Hamel 2002; Hamel and Prahalad 1994; Kelley 2001; Maxwell 2007;
Prahalad and Krishnan 2008). These concepts gained credibility when
Google, Facebook (now Meta), and Amazon rose to the ranks of industrial
giants. Roberto Verganti's research brought these concepts forcefully into
design discourse in papers that later led to *Design-Driven Innovation* (2009).
The idea of radical innovation ran counter to the idea of design as interpretation, but as we have seen, the empathic group was able to develop many
methods to respond to the call for radical innovation.

The shift raised another problem, however: the question of aesthetics. How should design look and feel if it is working under the auspices of
radical innovation? How should these radical ideas manifest themselves in
design outcomes in the context of interpretation?

The group knew that it had to develop an approach to aesthetics, so it
examined the research literature to find precedents. Three leading candidates that dominated the aesthetic discussion at that time were Italian radical designers, including Alessandro Mendini, Ettore Sotsass Jr., and Andrea
Branzi. These designers' aesthetic differed radically from commercial aesthetics and critical designers, including Anthony Dunne and Fiona Raby,
whose designs were intentionally so disturbing that they could not be mis-
taken for products (Dunne and Raby 2001, 2013; Dunne 2005). In human-
computer interaction (HCI), Genevieve Bell and her colleagues had argued
that designers should make things strange by defamiliarizing them (Bell,
Blythe, and Sengers 2005). These approaches aligned design with twentieth-
century avant-garde art rather than the industrial MAYA (standing for "most
advanced, yet acceptable") adage of Raymond Loewy, one of the founders
of industrial design (e.g., see Loewy 2002).

These approaches presented a problem for the group. Rich as they were
in expressive power, they did not fit the interpretive beliefs of the group.
They were consistent with the idea that designers are legislators, but the
idea that designers are in the aesthetic vanguard did not suit the idea that
designers ought to be interpreters who help people to understand their
peers and things in their world (for legislators and interpreters, see chapter 1). The group's problem was how to find an approach to aesthetics that
would respect its beliefs.

A more satisfactory approach took shape in stages. The literature offered some precursors in the search for an interpretive approach. One was the pragmatic aesthetic of Graves Petersen and her colleagues. It was abstract but it turned aesthetics into an empirical problem by situating it into "the social-cultural context of people's everyday life" (Graves Petersen et al. 2004, 275). Exhibitions provided other potential precursors. For example, Naoto Fukusawa and Jasper Morrison's *Super Normal* (2007) and Konstantin Grcic's *Design Real* (2010) had celebrated the aesthetic of ordinary, industrially produced objects (which was typical of any industrial design exhibition).

The group's experience offered more precursors. Most of its members were designers who had learned to find aesthetics in the studio, a bit like actors who find the right expressions before the camera turns on or the performance begins by experimenting and rehearsing their lines. Finally, the group's research had some precursors as well, including Turkka Keinonen's industrial work at the Nokia Research Center (Korhonen 2000, 190); Tuuli Mattelmäki's probe studies, which brought the group into contact with critical design; Andrea and Marcelo Júdice's *Vila Rosario* project, which captured the indigenous aesthetic of the village (Júdice et al. 2015); the theses of Maarit Mäkelä (2003) and Kärt Summatavet (2005), which showed how skilled craft could be turned into a research instrument; and Mattelmäki's visit to Pasadena in 2007, which introduced her to the complex aesthetic landscape of the ArtCenter College of Design and Los Angeles.

Mattelmäki's work was particularly important. When back in Helsinki, she started to use design techniques like exhibitions as research techniques. The first place where she experimented with these ideas was the *SPICE* project. Her new methods suited the empathic program in at least two ways. First, they complemented the old palette of design methods. Second, her new methods complemented books and conference papers as methods of distribution. The *SPICE* and *eXtreme Design* projects showed that these methods also worked in industry, which was used to trade shows, exhibitions, and fairs (Johansson et al. 2010). Her influence was felt in many student projects and master's theses that turned into exhibitions and used metaphors liberally. The first doctoral studies that explored these themes were published in the early 2010s (cf. Vaajakallio 2012), and the first doctoral theses that were driven by the new spirit were Acharya's *Opening the Electrome* (2016) and Wu's *Bicycles and Plants* (2017). The latter was also the first piece that experimented with design fiction, another thread of research that originated in Los Angeles.

A broader answer came from the group's experience at a respected design school. Seen from within the empathic group, the claim that radical innovation had to look and feel strange was premature in the end. Except for Acharya, this idea was far from how most empathic designers saw goodness in design. The group's approach to industrial design was professional and contemporary. They could also look at the craft world to see how it was using art to convey its ideas to the public. Their colleagues in craft explored design through art, but even when their design pieces were exquisite, few of these were avant-garde and disturbing. For example, Maarit Mäkelä (2003) created beautiful, semitransparent ceramic plates that explored human relations in her family, Kärt Summatavet (2005) created skilled brooches based on her study of Estonian folklore, and Elina Sorainen (2006) explored ways to revive a 4,000-year ceramic tradition by setting up a living museum at Kalporagan, Iran. These designs were magnificent, but they were neither shocking nor even mildly disturbing. Yet many of them were years ahead of their time. Mäkelä's autofictive and retrospective methods remain unmatched in design research, as does Sorainen's work in Persia. The baseline for an aesthetic was safe. In defining the aesthetic, the group could rely on its design training, which had proved its value again and again.

But this was the baseline rather than a theoretical answer. A theoretically reflected answer to the aesthetic question eventually started to take shape in conversations between Katja Soini and me (for more on her work, see chapter 3). When the latter started her doctoral studies, she took my research methods class. From that class, I knew that she was interested in conversations and dialogue, so I suggested that she read Grant Kester's texts about conversational and collaborative art (Kester 2004). She had been a top student in high school, and she took the challenge. After a year of reading and thinking, however, she concluded that conversation is a solid approach, but it was not right for her purposes. By exploring art, she discovered her identity as a designer.

A few years later, I built on these conversations when I explored the aesthetics of social form in social design (Koskinen 2016). I concluded that it is better to detach the idea of radical innovation altogether from how design looks and what it feels like. Design does not have to look radical to *be* radical. Instead, it can follow conceptual art that since the 1960s has recycled and reorganized existing forms, in order to direct the attention of the audience to conceptual ideas rather than artists and their skills. This realization was consistent with the interpretive beliefs of the group. If design achieved its purpose, it did not need a distinct aesthetic. Its base could be the extension of a subway line in Helsinki, a meter hack in Karnataka, a fictional plant hotel in Korea, or a living museum in Persia. It could also be a name-card holder, a ceramic urn for deceased family members, a joystick used to control a forest tractor, or a wrist-top computer for surfers.

This approach to aesthetics was flexible. It directed the group to work with the aesthetic beliefs of those people whom researchers were studying. It meant that in an industrial context, the aesthetic could capture the prevailing aesthetic of the time and build on it, following Loewy (2002). In a design studio, it could be the leading design language of the time. In an artistic context, it could be Dunne and Raby's disturbing and speculative avant-garde style. And in a local community, it should speak to people of the community. As Katja Soini and Heidi Paavilainen noted in the *Ave Mellunkylä!* project, "Mellunkylä inhabitants do care about their environment and its aesthetics, but the design needs to become personally meaningful" (2013, 93; see also chapter 3 of this book). If designers see themselves as interpreters, perhaps they could be midwives in a process that builds, elaborates, and extends aesthetics in its natural environment. If this is the case, there is no silver bullet for aesthetics; it is material that design researchers need to capture with their professional skills in every project to find a working expression. Why disturb people if you can achieve your ends by working with them, by capturing their sense of what is good? Why should we couple aesthetics with theory and midcentury avant-garde when we can study it in normal and real things like scissors, industrial robots, table manners, gift-giving rituals, and travel habits?

Smart Products

Usability of Smart Products

Maypole

One-Dimensional Usability

The Future of Digital Imaging

Cardboard Mock-ups

eDesign

Design Games

IKE

Empathic Design

Luotain

Design Studio in the Field

Mobile Image

Co-experience

Collaborative Design

Racing...

Design Probes

Prototyping Social Action

Women and Jewelry

you are Important!!

Design for Hope

Dwelling with Design

Morphome

Electronica

Pasadena

IP08

Art of Research

Prototyping Interactions

SRILE Project

Design Research through Practice

Bikes and Plants

Against Method

Memories in Clair

Drifting by Intention

Folk Tradition

Ryijyriuja

What Happened

Novapro

From Disposable to Sustainable

DWoC

Lost in Woods

5 Interpretation and the Constructive Turn

This chapter describes how technology can be integrated into an interpretive foundation. This foundation gave the empathic group tools to study user experience, but there was a catch. These methods and frameworks pushed the group away from technology. Design, however, not only imagines but also builds things that have not existed before. It produces things and products rather than descriptions and explanations.

To address this problem, the group examined design as a research tool and hired practitioners and engineers. The *Cardboard Mock-ups and Conversations, eDesign, Maypole* and *Empathic Design* projects repurposed design-based sketching and prototyping into research. *Morphome* project explored interactive devices to study proactive information technology. One of its outcomes was an interactive laboratory that built prototypes.

These explorations later turned into a prototyping methodology in *Prototyping Interactions*, constructivist methodology in *Design Research through Practice,* and a study of its philosophical foundations in *Drifting by Intention*. These studies proved that an interpretive approach can lead to physical and interactive prototypes.

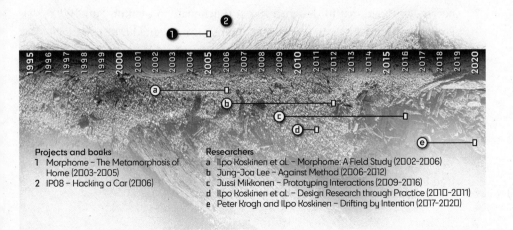

Projects and books
1 Morphome – The Metamorphosis of Home (2003-2005)
2 IP08 – Hacking a Car (2006)

Researchers
a Ilpo Koskinen et al. – Morphome: A Field Study (2002-2006)
b Jung-Joo Lee – Against Method (2006-2012)
c Jussi Mikkonen – Prototyping Interactions (2009-2016)
d Ilpo Koskinen et al. – Design Research through Practice (2010-2011)
e Peter Krogh and Ilpo Koskinen – Drifting by Intention (2017-2020)

The interpretive foundation was a treasure trove to the empathic group. It gave the group methods for studying how users create their experiences. It made research more precise, especially in areas where contextual variables matter, including mobile technologies and other smart devices. It also avoided the need to learn complicated theories and helped researchers to connect with many senior designers, whose mission was to interpret their environment and turn it into products and services. Merging the foundation with codesign and imaginative techniques gave the empathic group new abilities and extended its conceptual reach.

These changes came with a price, however. The shift in foundation pushed design further from technology. Yet, technology is crucially important for any design research program worth its salt: designers are accountable for imagining and building things. Its instruments are products, graphics, interactive devices, spaces, and systems rather than descriptions and explanations (e.g., Buchanan 2001). It is also a key part of international definitions of industrial design that see it as a discipline between technology and society (WDO 2022). Instead of pushing designers into the studio, the interpretive foundation encouraged them to go out into the world to study people. This move created new audiences for design researchers, but at a price: it also created a distance from the technological side of design.

How to do research that leads to things was becoming a hot topic internationally after the turn of the twenty-first century. In the 1990s, MIT's Media Lab had twisted the old academic proverb "Publish or perish" into "Prototype or perish" (Negroponte 1995), human-computer interaction (HCI) had developed a nascent literature about prototyping (Säde 2001), and the Designing Interactive Systems conference had published several design papers that became foundational in design research (Buchenau and Fulton Suri 2000; Forlizzi and Ford 2000). In Los Angeles, Brenda Laurel edited *Design Research* (2003), which described some methods used at the ArtCenter College of Design in Pasadena. In Ohio, Liz Sanders (2000) talked about generative methods. In London, Dunne and Gaver's *Presence Project* (2001) built interactive benches and deployed them in Norway, Italy, and the Netherlands to study how people use them. In the Netherlands, Stephan Wensveen (2004) and Joep Frens (2006) showed that prototypes can be research instruments. Soon afterward, the notion of research through design started to spread rapidly in American interaction design after the publication of a paper by John Zimmerman,

Jodi Forlizzi, and Shelley Evenson (2007). Moreover, there was an emerging craft research community in northwestern Europe, and critical design started to gain popularity in European design schools (Dunne and Raby 2001).

Some references were closer to home. In Scandinavia, Pelle Ehn (1998) made a call for a digital Bauhaus, participatory design researchers in Aarhus were creating prototypes in a research framework with considerable success (Lykke-Olesen 2006; Ludvigsen 2006), and the Nordic design research community started to meet at the Nordes.org conference series in 2005. The empathic group also felt the pressure to tackle the question of making from within itself. It was on top of the European Union's *Maypole* project, which built prototypes of mobile multimedia phones (Mäkelä et al. 2000). The notion of "practice-led research" landed in Helsinki through craft (Mäkelä 2003; Mäkelä and Routarinne 2007), and several theses combined roots in the studio with fieldwork (Säde 2001; Summatavet 2005; Sorainen 2006). The group has explored action research and was aware of many research fields built on practice, including clinical medicine, constructive accounting, and organizational development. The confluence of these references led to what the group started to call a constructive turn in design research (Koskinen et al. 2011).

This chapter describes the group's ten-year path from interpretive to constructive research. It tells the history of how the group came to answer the call to make things, how it made its interpretive foundation elastic enough to accommodate even electronics, and how it came to integrate making into its theoretical foundations. The main ingredients were traditional three-dimensional (3D) design techniques, collaborations with electrical engineers and computer scientists, and literature about prototyping in user-centered design. As the chapter shows, the empathic group was well resourced to take up the challenge. It had its origins in a resilient design community that had learned to value ergonomics and usability research—in the 1960s. Yet, even though the community accepted the research and found value in user-centered and empathic methods, it raised a number of questions about design skills at the base of the discipline. Where was the stuff? How could design skills be used in research? What should be the role of prototyping in research? These questions were legitimate. As any designer knows, ideas may sound good and look good in a rendering but not feel right when you actually touch and try them.

Cardboard Mock-ups and Conversations

The first consistent effort to bring construction to the group's research at a doctoral level was *Cardboard Mock-ups and Conversations,* a project by Simo Säde (2001). He was working with ED-Design, the country's leading industrial design consultancy at that time. His goal was to find and introduce user-centered design techniques to the company. Säde's approach was loosely based on Kurt Lewin's (1946) action research. His study was a partial success. One of the partners of the company became his sponsor and the project's user-centered leader, but it was difficult to sell new methods to other partners. His conclusion was that he had created a targeted toolbox of methods for the company rather than changing it profoundly. The toolbox metaphor was apt. As in any toolbox, some tools are used often, while others are just waiting for those rare moments when they are needed.

The most important consequence of his thesis was that it showed to the rest of the group that design techniques can be turned into research methods. His literature review was particularly useful. It classified prototypes from low-fidelity drawings to high-fidelity interactive and realistic prototypes. The idea of subsuming design under the concept of prototyping was popular in the 1990s and critiqued only later by authors like Buxton (2007), who made a sharp distinction between sketching and prototyping. One main benefit of Säde (2001)'s review was that it did not see prototypes in an industrial fashion as the first in a series or a preparatory step for manufacturing. Techniques like cardboard modeling and paper prototyping of user interfaces were scarcely prototypes in the traditional industrial sense. His study showed to the rest of the group that many concepts that industry took for granted were in fact conventions that had to be redefined in the research context.

He used both field-based and studio-based methods. In one study, he built a mock-up of a user interface of a forest tractor and a user-centered method toolkit. The mock-up was built into a studio with a one-way mirror for observations. To explore the controls of the tractor, he built paper prototypes and mock-ups of these controls. In this environment, he conducted usability studies with Battarbee, who helped him to conduct the tests and videotape the conversations. Next, he created interactive models of the interfaces in Macromedia Director. In another study, he built virtual prototypes with computer scientists to test whether they could be substitute mock-ups and real prototypes. He produced a Virtual Reality Modeling Language (VRML) model of a prototype, and for comparison, a 3D print of the model that was sanded and painted to give it a finished feel. This work was designed for a consortium of several mobile phone and personal electronics companies that were interested in whether they could move their usability studies online.

Importantly for the future of the group, Säde showed that mock-ups and prototypes can be studied in the field in real-life contexts. In one study, he built cardboard mock-ups of a bottle recycling machine for Halton, a manufacturer of bottle refunding machines for grocery stores. He built two cardboard mock-ups, one in which bottles and cans were put into the machine vertically (with the top up) and another in which they were entered horizontally (with the bottom on the left). He tested these two designs in groceries with a 1:1 size cardboard mock-up. The behaviors of the machine were mimicked by a colleague sitting behind the mock-up, pretending to be the mechanism and mimicking its behavior to shoppers recruited from the grocery. The results were clear: the vertical model was clearly better, and it went to manufacturing. (Säde et al. 1998).

The theoretical purpose of Säde's thesis was to clarify discussions about prototyping in design research. He mapped a wide range of prototyping practices and classified them into the popular cline from low-fidelity prototypes to high-fidelity prototypes. This reflected the reality of his research, which ranged from rough cardboard mock-ups in the Halton study to high-fidelity models in the VRML study. The distinction between low- and high-fidelity prototyping gave a scale to his taxonomy. It served him well in creating the methodic toolbox for ED-Design. His techniques at the low-fidelity end were timely: many leading researchers at the time were expanding the definition of prototyping to find cheaper, research-based alternatives to it. What was lost in depth was gained in breadth, but his thesis exposed his coworkers to a wide range of design-based research practices and gave them a precedent for turning prototyping into a research tool. He also extended the notion of prototyping from physical devices to conversations.

After Säde finished his thesis and moved to industry, the rest of the group kept his focus on the front end of product development and picked up the field-based strand of his work, but it also shifted its focus from interfaces to people. Tuuli Mattelmäki sought ways to study people for inspiration, Katja Battarbee ways to capture co-experience, and Esko Kurvinen ways to prototype social action. The program's shift to developing methods and frameworks for capturing user experience served the group well, but it also pushed it away from doing design. High-fidelity prototyping was put on the back burner for a couple of years.

Morphome: The Beginnings of Constructive Design Research

The shift to a low-fidelity approach was academically fruitful, but in the back of the minds of the researchers was a nagging question. In the words of Raimo Nikkanen, the head of the industrial design department at that time and a veteran designer, empathic design was focusing on predesign. When listening to him and his senior colleagues from industry, the group saw that they were happy that design was finally producing credible research that was also close to their occupational interests. However, they also wanted to see how predesign feeds into design proper. Back then, the billable hours in the design industry started from the moment that designers started to work in the studio. Predesign, in contrast, was mostly presales: it was possible to sell a few interviews and observations to industry, but it was hard to put a price tag on these hours. How could the empathic agenda be brought closer to the world of billable design?

The first step toward an answer came from a collaboration with the humanities and electrical engineering. Frans Mäyrä, a literature scholar, game designer, and talented young professor of hypermedia at the University of Tampere, Finland, was interested in proactivity, an avant-gardist concept in information technology (Mäyrä et al. 2006). This term had been introduced a few years earlier by David Tennenhouse (2000), who presented a vision of information technology that could use information from sensors to predict human action and act before people become aware of their needs. This in turn reduces the cognitive load that technology imposes on us. If technology acted without human intervention, people would not need to pay attention to it. Technology would become calm.

This was an attractive vision then, and it would be even more attractive today, when social media has turned most of us into click-a-holics. For Mäyrä, Tennenhouse's vision of calm technology raised fascinating questions about context. The best-known example of proactive technology at that time was antilock brakes in cars, which had made them safer. That development proved that it was possible to build proactive technologies that targeted well-defined problems. How would the vision work in a larger context of human activity like the home, he asked? What kinds of design principles should we follow when designing proactive technologies for the home, an intimate place for relaxation and family interaction?

Mäyrä contacted electrical engineers of the Personal Technology group, led by Jukka Vanhala at the Tampere University of Technology. He also knew that to conduct research about proactive technologies, he needed designers. He was aware of the multimedia studies that I had done with Kurvinen and sent an email to me. Kurvinen was busy with other projects, but Battarbee was looking for a final project to wrap up her doctoral thesis. The project was named *Morphome—Metamorphosis of the Home* (Mäyrä et al. 2006; Koskinen et al. 2006). It started with user studies that helped the research team create visions of how proactive technology would work in the home. It went on to prototype proactive products and how they would communicate with each other over a network. These prototypes made it possible to study what kinds of proactive technologies (if any) people would be willing to have at home.

Morphome quickly discovered limits that most smart home studies had seen earlier (Harper 2003). Some limits were industrial. It was impossible to network existing appliances like vacuum cleaners with lights because of intellectual property issues and a lack of technical standards. There were ways around these limitations, of course. The project built rough prototypes to give people the experience of living with devices that talk to each other and make independent decisions, but they faced constant technical difficulties. In 2003, one problem was battery life, which set limits to communication and needed complex communication management. Some difficulties were relatively easy to overcome. For example, the project interlinked lights and emulated a proactive home in which several devices would interact without human intervention with the X10 smart home system. The main difficulties were ethical, however. People repeatedly expressed concerns about having these technologies at home, and many of them referred to Stanley Kubrick's famous film *2001: Space Odyssey,* with its apocalyptic HAL scenario. For them, the home was for relaxation, recharging, and fun, not for producing measurable output or controlling the environment. They also raised concerns about privacy. Who would have the right to follow the stream of information from these devices? Would having proactive technology at home give companies and governments a window into their private lives?

These limitations aside, the study cleared some paths methodologically. Its combination of construction of technology and fieldwork was advanced at the time, and it also had quasi-experimental overtones (Koskinen et al. 2006). In *Morphome*, fieldwork and construction went through three cycles: a pilot study with interactive cushions, a study with proactive lights, and an X10 study that mimicked a proactive home. In all, more than sixty users took part in these studies, which took two years to finish. Over the course of the project, the research team grew more skeptical about the proactive vision. While proactive technology was working perfectly well in clearly specified contexts like antilock brakes in cars, everyday life comes in many forms, with ambiguous boundaries. It was too difficult to build systems that could foresee social activities. *Morphome* became a critique of pushing proactivity too far into everyday life.

For the empathic group, *Morphome* proved the value of prototyping technology and combining prototyping with fieldwork. Although its methodology was intensive and slow in terms of prototyping design concepts, this was the first major empathic study that prototyped complicated technological concepts seriously, and it clearly demonstrated the tensions between field research and prototyping. Over the next few years, the project led to a better integration of these two activities. The goal of this integration was to give prototyping more room in design research.

Hacking a Car: Interpretive Foundations in Electronics

Morphome, which was over by 2006, combined an interpretive approach to people and electronical prototyping. In 2006, however, it was not clear how that combination would work in methodological terms. Specifically, the missing part was the role of prototyping in a larger interpretive framework. The idea that prototypes have a methodological function was in the air, however. For example, Caroline Hummels talked about prototypes as physical hypotheses in her thesis (Hummels 2000), and the idea that prototypes are provocations and interventions was explored by Preben Mogensen and Genevieve Bell (Mogensen 1992; Bell, Blythe, and Sengers 2005). How would prototypes function in an interpretive framework?

There were some indications in the research literature (e.g., Crabtree 2004), but the answer to this question took a long time to find. The idea of seeing prototyping as a methodic device came to my mind in December 2006 in

Pasadena as I looked at the snowy mountains in the horizon over Altadena. I realized that the literature at that point clearly fell into three categories. First, there was excellent recent experimental research from the Netherlands (e.g., Wensveen 2004; Frens 2006). Second, there was highly sophisticated, fieldwork-based research coming from the Nordic countries and from companies like IDEO in Silicon Valley (Ehn 1988; Buchenau and Fulton Suri 2000; Koskinen et al. 2006). Third, there was research that built on design practices that did not fit into either of these categories, including Italian *controdesign* and critical design (Dunne and Raby 2001; Bell et al. 2005; see also Branzi 1988). There was also scientific literature in business technology, but I have discounted it because of its disciplinary foundation was distant from design (see Hevner et al. 2004). I realized there was enough literature to support a consistent methodological argument, but it had not been compiled into one coherent thesis.

When I returned to Helsinki in 2007, I collected the literature about how design could be turned from practice into research and put it into the three areas that I had sketched out during my lunch. I wanted to push user-centered design into what I came to call a constructive methodology. To see how this idea would work, I initiated Interactive Prototyping, a class for prototyping interactive technology. The department hired Jussi Mikkonen, an electronical engineer who had taken part in *Morphome*, to teach and oversee the technical part of the class.

After piloting the class in 2007 with doctoral students, the first real implementation was in 2008 (IP08 2009). The topic of the class came from government statistics showing that two of the most dangerous activities that drivers engage in are talking to and feeding children in the back seat (Summala et al. 2003). The students were to create concepts that make these unavoidable interactions less dangerous. The framework of the class came from Battarbee's thesis: the students were to study "co-experience in the car" and build their concepts and prototypes around their observations. How could one create designs that would make these interactions fun and safe to do while driving?

In terms of technology, the students had to learn the basics of a microcontroller (ATmega8535) and elementary programming in the C language and refresh their knowledge of the basics of electronic circuits. The technology partner of the class was VTI Technologies, which supplied sensors to most auto makers and was among the world leaders in its field. Its

Figure 5.1
Hacking a car. A BMW was bought for the IP08 course and driven into one of the computer-aided design/computer-aided manufacturing (CAD/CAM) studios at the university. Here, two students are working on their concepts. (Picture credit: Ilpo Koskinen, 2008.)

headquarters was in the northern suburbs of Helsinki. The students were not given any design guidelines, other than that they had to build functioning prototypes that could be tested at the very end of the class. An old 1989 BMW was bought for the class. It was registered and in full driving condition, but it was also old enough to be deconstructed in the studio space devoted for the class. Being in a studio meant that students working on the car had easy access to equipment and fellow students who could be recruited for user tests (figure 5.1).

Day 1 of the class included introductory lectures, a lecture from a physicist from VTI Technologies, and a session in which students had to play with a Nintendo Wii to experience firsthand how a motion sensor reads their movements and translates them into actions on a screen. The first piece of homework for the students was a user study that was to be conducted during the first week of the class. The students were instructed to interview and photograph one or two families with children and at least one car. They had to conduct the interviews in the car, document interactions in the car through design-specific research techniques, ask people to act out typical interaction situations, photograph or film these situations, assume the seating positions themselves to experience them firsthand, play these situations by themselves to get feedback from interviewees, and take measurements to understand ergonomics and dimensions. They were finally instructed to do all this from both the parents' and the children's perspectives to internalize the possibilities of the interior.

After the user study, students had to go into the studio and act out these situations in the car to understand and to internalize them from multiple perspectives. Students had to explore their findings by acting out stories that had been told to them, by role-playing them, and by bodystorming (i.e., physically situated brainstorming; see Buchenau and Fulton Suri 2000). After bodystorming, they had to develop three concepts based on their findings and explore them in the car. Specifically, they had to create a low-fidelity prototype and play it from the driver's and the backseat passengers' perspectives. When they had a leading concept, they had to sketch it in the car with whatever means they could invent. These means ranged from interaction sketches executed with Lego Mindstorms to experience prototyping screen-based communication between the parents and the children with video cameras. After these role-plays, they abandoned some ideas because bodystorming showed that they did not feel good. For example, one team gave up the idea of using screens to mediate communication between the front and the back of the car. The resolution was too low, and placing what was essentially a television or computer screen in the backseat was neither novel nor engaging.

Some concepts fared much better. For example, one concept turned the body of the car into a musical instrument. Two students discovered that children often tap various parts of the car when listening to music, or even without music. Their idea was to place sensors in the car to give these little drummers a musical instrument. These students bought cheap toy instruments, broke them down to understand how they worked, and rewired the existing stereo system of the car with components scavenged from the teardown so it was possible to channel tapping into the car's entertainment system. The same pair also turned the rear-view mirror into an instrument that the driver could use to give an angry look at children when they started to fight or got too loud. Over the course of the prototyping, the students realized that the latter concept could not be fully automatic. Most of the problems were related to how to make movements of the mirror smooth and timings intuitive. Instead, they built a control button onto the steering wheel.

One trio observed that children want to know what is happening outside the car and are eager to know the remaining time to the destination ("Are we there yet?"). They turned the observations into a game that gave children an opportunity to experience the road without the driver's active input. They hacked the steering system and the car's speedometer to create a game that used information from the steering system to build a skylike game involving "bugs" into the ceiling over the back seat. These bugs were small, plastic, yo-yo-shaped cones with light-emitting diodes (LEDs) and proximity and touch sensors. When one of the bugs lit up, a player got points if she managed to turn it off by waving her hand in its proximity. Mistakes led to losing points, as did touching the bugs. A game engine calculated the result, and a screen showed all the results after five minutes. It was a one-way interactive game that reacted to road conditions. Children could interact with the driver and the road, but the driver was not disturbed by the game.

Over the new few years, the Interactive Prototyping course and Mikkonen's laboratory created dozens of prototypes, which Mikkonen lists on his website, kryt.fi. One class of prototypes consisted of augmented video images, including projects like *Honest Shoes* (Kwak, Suomalainen, and Mikkonen 2011). Other projects included *Hello Bracelets* and *Spatial Jewelry* (Ahde and Mikkonen 2008; Ahde, Mikkonen, and Latva-Ranta 2009), *OJAS* (Mikkonen et al. 2014), *Light Is History* (Acharya, Bhowmik, and Mikkonen 2013), and *Butterfly Lace* (Kuusk, Kooroshnia, and Mikkonen 2015). All these focused on augmenting existing products with sensors that could "see" their environment and act based on what they found. Petra Ahde-Deal's *Hello Bracelets*, for instance, tried to sense emotions and communicate them to other bracelets in the system. It was inspired by *Bling Bling*, her master's thesis (for more about this, see chapter 2). Another class of prototypes consisted of wearables, Mikkonen's technical specialty (Mikkonen and Townsend 2019). He worked on a heating jacket for outdoor workers like loggers and construction workers, for example.

He also helped students to build concepts for Bang & Olufsen in two projects, *BeoSound Orbit* and *BeoSphere*, and worked for Microsoft on the *Domesticating Search* project (Sellen et al. 2011), and for Nokia on several student projects. His work has led to several start-ups and patents, some developed from design sprints that he has been running. The work of the studio has been exhibited widely, such as in Belgium and Germany and at Shanghai's World Expo in 2010. Mikkonen also took part in several doctoral theses as a technical advisor and helped the Berlin University of the Arts to start interactive prototyping. At least two designers from the 2007–2008 classes became principals in their own design firms in Berlin and Shanghai.

The achievements of this class proved that interpretive design research methods work well with open-minded engineers. The foundation of the class was laid in 2007, when Mikkonen and I decided to keep its research phase short and directed most of our efforts toward prototyping. Another crucial decision was to build the course on fieldwork. The design instructors of the course have changed several times over the years (this rotation was planned), but most of them have taught the students to start with fieldwork. Mikkonen's technical and imaginative skills aside, interpretive fieldwork was the glue that kept the class together until 2017, when Mikkonen left the university.

Prototyping Interactions: **A Taxonomy of Prototypes**

In addition to Interactive Prototyping itself, Mikkonen was tasked with building a laboratory to support the class and encourage design research to complement the design school's workshops and 3D facilities. The laboratory that he built was small, but its effects on design research went far beyond the limits of the physical space. Its most crucial resource was the students. Even though the number of students was limited, the studio produced more than ten prototypes annually, some by students and others by researchers. Some of its alumni became remarkably successful designers. They had a method, they could describe it to their customers, and they knew how to work under time pressure and turn stress into an asset. Having the ability to turn a driver's anger into a piece of embedded communication technology is a skill that any good designer can appreciate.

The laboratory had a more ambitious goal as well. It was to find ways to instruct students and the technologically capable members of the empathic group to think in terms of a machine. Knowledge of the way that an electrical system senses its environment and processes its findings into actions is a skill that some designers needed. This skill gave them an insight into how engineers work, which improved their communications with industry. The studio believed from the beginning that it was important to know what users need, but this was not enough to build interactive systems. The laboratory was asking how the behavior of the prototype affects users and the design process, how it changes the context of use, and how this change could be reflected in later iterations of the prototype. It shifted the focus of the empathic program away from humans and toward understanding technological systems, how humans interact with these systems, and how these interactions could be fed back into design.

The laboratory soon proved that its methods were working. The interpretive language behind it, however, posed some problems. Interpretation was not the language of industry. When dealing with it, Mikkonen had to use its terminology. In his doctoral thesis, *Prototyping Interactions* (2016), he set out to develop a classification that would work in industrial collaborations. He took a taxonomical approach to prototyping and built his classification around the purpose of a prototype. For him, prototypes can illustrate functioning design ideas, illustrate rich and tangible interaction concepts, explore body-based and social actions, and examine vague ideas in the sketching phase. He built a five-class taxonomy to capture his experience, which merged prototyping literature and user studies. The first two bullets support research, the last three product development:

- *Ideation sketches.* These are typically simple, quickly done, and have a shorter lifespan than prototypes that support user studies.
- *Supporters of user studies.* The purpose of these prototypes is to develop ideas further by introducing them to users. These are research-led and unproductlike, and their purpose is to gather information. These two types can also be called "research prototypes."
- *Proofs of concept.* These prototypes are implementations of innovative ideas for the first time, and they are done when ideas mature to patent applications and to justify costs.
- *User-proof prototypes.* These are a step up from proofs of concept. In the world of engineering, they are built to test features numerically, while in the design world, they exist to test forms, aesthetics, and user experiences.

• *Prototyping enablers.* Prototyping enablers are systems needed to enable the creation of other devices. In the studio, this involved a capacity to build microcontrollers from a scratch, but also packages like Lego Mindstorms, Arduino, and Raspberry Pi, as well as software compilers. An example of Mikkonen's own work is the *OJAS* project, which developed an open-source, bidirectional inductive power link that illustrates the concept. It became the basis of a start-up and numerous prototypes by master's students (Mikkonen 2016, 42).

Mikkonen's thesis provided a much-needed clarification of prototyping from the perspective of an engineer with ample experience in working with interpretive designers. His classification was discrete; his five types were different, but they were not ordered. His terminology was better suited to his studio environment than Säde (2001)'s scale from low fidelity to high fidelity or Buxton (2007)'s distinction between sketching and prototyping. Both lost the nuance that Mikkonen needed in his work not only with designers, artists, and researchers, but also with engineers, industry, and start-ups. In his environment, he had to be able to quickly tell fellow designers, students, and visitors what was relevant for them, and he also needed a well-grounded schema for his own work. His classification worked well for both purposes.

The user-centered aspects of his taxonomy were informed by his experience in one of the world's most research-intensive departments of industrial design. This was evident in the notion of "supporting user studies," which had been missing in Säde's thesis and its background literature. Mikkonen's taxonomy was nuanced and flexible, and it merged the human and technical sides of design. It stood firmly in line with empathic design in its focus on industrial products. His approach was neither critical nor utopian; it was research-based and grounded in experience. He claimed that his taxonomy offers more sensitivity toward the act of prototyping than the taxonomy of Wensveen and Matthews (2015), and he may well have been correct (Mikkonen 2016, 46).

Toward Methodology: *Design Research through Practice*

Around 2007, the literature about prototyping was mostly grounded in industrial language, but it did not reveal what kind of vehicle of knowledge design can be. The Interactive Prototyping class soon led to a methodological study that provided an explication of prototypes from a methodological perspective. A few months earlier in Pasadena, I had sketched out three methodological approaches to prototyping in design research. Specifically, I had wanted to understand how design pieces can create knowledge. I was not satisfied with the taxonomical approach that had dominated the discussion of prototyping for a decade and wanted to avoid the dichotomy between science and design that had dominated a good deal of design literature. I thought it was wrong and left major gaps in discussion. My remedy was a sketch for a methodological approach.

I first published the three approaches in the sketch as *Lab, Field, Gallery and Beyond,* with Thomas Binder and Johan Redström (Koskinen, Binder, and Redström 2008). The lab approach in this paper came from Dutch work led by Kees Overbeeke in Eindhoven. In this approach, design pieces are physical hypotheses (Stappers 2007). Researchers conduct research and create concepts that they turn into prototypes. As soon as the prototypes are functioning, they can be tested through experimental methods. Testing is done statistically. The best example we could think of was Joep Frens (2006)'s doctoral thesis. Building on J. J. Gibson's ecological psychology, Frens had built a rich interaction for a digital camera. By "rich interaction," he meant an interface that would use natural gestures rather than drop-down menus as the control. He built four camera variations—one rich, one traditional, and two in between. He tested them with students of architecture in an experimental setting and analyzed his data statistically. He worked like a scientist testing a hypothesis in a laboratory.

This was markedly different from the field approach. Our example was the *Morphome* project, in which the team had built proactive devices and linked them to a network. Instead of testing them by varying them in an experiment, the *Morphome* team gave the prototypes to several households to observe what would happen to them. Gone was an attempt to control the variables that shaped the uptake of technology, and gone was the idea of testing. Instead, the study followed what happened to the prototypes. The researchers' job was to carefully describe the life of the prototype—in brief, work like what an anthropologist would have done. The gallery approach was an amalgam of experience from the ArtCenter's Super Studio and Dunne and Gaver's *Presence Project* (2001), which had introduced the idea of cultural probes to design (Gaver, Dunne, and Pacenti 1999). This approach gave up control and even the idea of careful observation. It saw user research as an inspiration and encouraged researchers to build concepts like design practitioners. It encouraged researchers to turn these concepts into prototypes imaginative enough to make people see that products around them are conventions rather than necessities. If it gave deployed prototypes for field testing, its manner of collecting data from these tests was deliberately anecdotal and placebolike, as in Dunne and Raby's *Design Noir* (2001).

The triptych form of the 2008 paper was soon expanded into *Design Research through Practice*, a book that I wrote with Binder, Redström, John Zimmerman, and Stephan Wensveen (Koskinen et al. 2011). In that book, the gallery approach of the earlier paper was called Showroom, using a term from *Design Noir*; this term anchored the approach in design rather than art. The book situated the three approaches historically, elaborated their theoretical commitments, studied the roles played by "design things," and speculated on the future of design research. Its method was explication, and the object of this explication was a few exemplary research design research programs, including empathic design, critical design, Italian work on sustainable design, and Nordic participatory design.

The book's core construct was the concept of constructive design research. This notion came from accounting (Kasanen, Lukka, and Siitonen 1993). Over the years, researchers have developed many ways to describe practice as research, but these were becoming confusing. For example, Richard Buchanan distinguished clinical research, applied research, and basic research (2001, 17–19), and Simo Säde (2001) talked about action research in his doctoral thesis. The notion of research through design, initially only a footnote, was going through a renaissance following the highly influential paper by Zimmerman, Forlizzi, and Evenson (2007) in the CHI conference. Constructive design research was a way of saying that construction is important and the key to design research of the sort that the book was interested in.

From an ethical perspective, *Design Research through Practice* was an argument for methodological tolerance. For several years, I had been intrigued about a pattern of speech that I kept hearing in design conferences. I heard presentations that contrasted science and design but found this deficient in two important ways. First, it equated science with experimentation, which is an inaccurate way of understanding research. As a former science student, I thought about research about bird migration and the formation of ridges and their geomorphology. There is room for experimentation in these fields, but the main methods of these sciences are not experimental. The same was true of theoretical astrophysics, in which progress happens through observation and theoretical speculation as much as through experimentation. I was also intrigued by the claim that experimental research is the only way to do science. For me, this was clearly too narrow, especially because the standard of experimentation was almost invariably adopted from textbook psychology. Accepting this claim would have meant that geology and theoretical astrophysics are not sciences, which is plainly absurd.

Another problem was that the distinction excluded the middle: it was black-and-white and missed an old debate about the differences between studying human beings and nature in philosophy, the humanities, and the social sciences. Disciplines like history and anthropology advance by collecting materials and putting them together into frameworks. These frameworks are treated as explanations in debate. This is how a good deal of the humanities works. Scholarship on Plato, William Shakespeare, and Johann Wolfgang von Goethe does progress, but more because of critical debates that lead to new insights than because of new findings. The sharp dichotomy between science and design allowed no room for interpretive knowledge in design. Still, a good deal of the best research in design was user-centered, and most designers thought that their discipline in fact mediated between human beings and technology. For me, this was a fact that had to find a methodological expression.

In the 2011 book, my coauthors and me wanted to break the either-or distinction between design and science in order to add tolerance to methodological debate in design research. It supplied a book-length methodological contextualization of interpretive research. Its motivation was firmly grounded in the experience of the empathic group and in work that had inspired it. Its scope was limited to methodology, however. It touched on the theoretical and philosophical underpinnings of constructive design research, but it did not go deeper.

The Epistemological Foundations of Construction: *Drifting by Intention*

One of the hallmarks of empathic design had been its proximity to design practice, but the main way that it propagated its ideas was participation in peer-reviewed research debates. *Design Research through Practice* focused on methodology rather than the concept of knowledge. By 2013, there was a growing body of literature about the methodology of design research, but its answers were varied and often even contradictory. Some saw no difference between design and research (Frayling 1993), others called for more methodic rigor to distinguish design and research (Zimmerman et al. 2007), and still others sought to link the accumulation of knowledge to research programs (Gaver 2012). The production of knowledge for the public domain is a characteristic that distinguished research from practice, but it was unclear how.

This question occupied Peter Krogh, my Danish colleague. He had an idea for a book about the epistemology of design research. He had published three papers that critiqued and expanded *Design Research through Practice*, but he needed to find a wider framework for a book. In our conversations, we realized that we might have enough material to study Krogh's epistemological interests through a body of more than sixty doctoral theses, many of which we had supervised and examined over the years. We decided to write a book about what Krogh had called "drifting," which we defined as those actions that take design away from its original brief or question and lead to a result that was not anticipated in the beginning. We believed that every designer—and every design researcher—drifts intentionally. The name of the book *Drifting by Intention* captured this belief (Krogh and Koskinen 2020; Krogh and Koskinen 2022).

The starting point of the book was our belief that earlier debates about knowledge in constructive design research were disjointed and overly abstract. We thought that a better starting point would be the idea that drifting is a function of the concept of knowledge. In the preface of the book, we told the story of Odo Fioravante, an Italian designer known for his chairs. Fioravante once gave a talk in Hong Kong. In the question-and-answer session afterward, he was challenged aggressively by one audience member about his disregard of scientific methods. His response was calm. He said that he was trained as a mechanical engineer, and he knows finite elements well, but experience is more important. By saying what every furniture designer knows, he showed that the challenge was built on a false assumption. Fioravante's response also shows how a theory of knowledge—that is, epistemology—affects design: the concept behind the challenge would lead to studying furniture through computer simulations, while the concept behind his historical approach would lead to observing classic chairs to see how fellow designers have solved problems with the chairs.

Krogh generalized this observation with me to identify the main ways in which constructive design researchers see knowledge and how their concept shapes their research. We articulated four concepts of knowledge in constructive design research:

- *Experiential tradition.* In the experiential tradition typical of traditional design in fields like furniture, knowledge is a personal matter. A designer can justify a shift in direction (drift) by referring to his or her intuition and history.

- *Methodic tradition.* In the methodic tradition, knowledge is equated with rigorous, usually scientific methods. In this tradition, drifting is justified if, according to a methodic procedure, the researcher's beliefs are wrong.

- *Programmatic tradition.* In the programmatic tradition, knowledge resides in discussions of frameworks like Andrea Júdice's *Design for Hope* (2014), which was indebted to Paulo Freire. Design pieces come and go and can be made sense of in many ways. Their meaning does not reside in the object: it resides in debate in the research community.

- *Dialectical tradition.* Finally, in the dialectical tradition typical of empathic and participatory design, drifting is a function of participation. In participatory design, for example, drifting is legitimate if participants say so and the researchers find no reason to doubt them.

Our book argued that if we look at diverse ways in which knowledge functions in design research, we can see that many of its debates are overly dramatized. In HCI circles, for example, many reviewers put John Zimmerman's and Bill Gaver's arguments about research into opposing camps and saw little similarities between them: science versus art, rigor versus creativity. Yet, although these arguments were seemingly contradictory, these researchers had no problems convening around empirical research and design pieces. For example, it was clear to Krogh and me that the programmatic tradition needs its design pieces, and although their meaning can change when discourse changes, they remain the foundation of research. The experiential tradition in its extreme prioritizes design researchers, but it can also lead to marvelous phenomenological studies of intuition, the body, or discussions in the studio—all potential sites of knowledge.

What would interpretive design research be like in the context of drifting? *Drifting by Intention* did not focus on interpretation, but it helped us to specify some of its philosophical underpinnings. First, interpretive design research had a natural fit with the dialectical tradition: when design builds on an interpretive intellectual foundation, design researchers study people to understand how they experience their world. Second, it was also compatible with the programmatic tradition. The frameworks that researchers build—for example, Katja Battarbee's co-experience—produce a conceptual form that can be discussed and revised by other researchers. If they survive debate and start to gain a following, they become precedents for subsequent studies. Third, it is harder to find room for interpretation in the methodic and experiential traditions. The experiential tradition does not require user involvement, and the methodic tradition does not require treating people as meaning-driven actors (although it can do so). These beliefs do not mesh well with the idea that the researchers' task is to interpret how people construe their experiences in interactions with others.

On a broader basis, *Drifting by Intention* wanted to add tolerance to design research, much as its predecessor had done in 2011. It was the first book that juxtaposed various strands of constructive design research to show that knowledge and design are linked in many ways. Each of these forms is well suited for some purposes and less so for other purposes. Interpretive design research sits comfortably in both the programmatic and dialectical epistemic traditions.

Interpretation and Constructive Design Research

One of the hot topics of design research in the 2000s was how to turn design into a research tool that produced knowledge, not just products, services, or concepts. Design practitioners were interested in learning how research could help improve design, and several research communities were exploring this topic. In its early years, the empathic group focused on finding ways to study experience and communicating its findings, and it shifted its balance to the human side. It knew how to study user experience, how to codesign, and how to complement its interpretive basis with imaginative methods that supported the generation of radical ideas, but these activities decoupled research from making things in the design studio. The group saw a need to offer something to practice or risk losing the design audience.

The answer to this conundrum was the concept of constructive design research. The topic of this chapter has been the emergence of this concept. The story started with the *Maypole* project (Hofmeester and de Charon de Saint Germain 1999) and the doctoral thesis of Simo Säde (2001), which tried to understand the changing notion of prototyping. The notion expanded significantly from the mid-1980s to the late 1990s. It moved the concept of prototyping from industry to research, and Säde brought the notion to Helsinki. Building on his work, a part of the group started to develop a methodological way to understand prototyping. A pivotal project was *Morphome*, a constructive study of proactive information technology (Mäyrä et al. 2006; Koskinen et al. 2006). It brought together field researchers, industrial designers, and electronical engineers. After finishing that project, the group hired Jussi Mikkonen, an engineer who had worked on *Morphome*. He created Interactive Prototyping, a class that taught prototyping to industrial design students. His class created dozens of prototypes and research papers, and his thesis, *Prototyping Interactions* (2016), created a framework for describing design prototyping to industry.

The shift to constructive design research was taking place in several design research communities simultaneously. It was evident in Scandinavia, the Netherlands, Britain, Italy, and the United States—and also in Helsinki, in projects like *Peukalo,* in which the industrial designer Teppo Vienamo studies hand ergonomics with 3D printed mock-ups. The unique contribution of the group was what it made from the shift. Namely, it created several frameworks that reconciled the worlds of interpretation and prototyping. For the Interactive Prototyping course, I created a scheme of three methodological approaches to prototyping. It built on the *Morphome* project, as well as my experience in Pasadena, but also on my knowledge of Bill Gaver's research in London and research in Delft and Eindhoven, among other places. The core concept behind Interactive Prototyping led to *Design Research through Practice*, a book that placed the shift in a methodological context (Koskinen et al. 2011).

Despite the shift toward constructive design research, design practice was never the main concern of the empathic group. The reason was probably the context in which the group was working. Most researchers in the group were designers by training. Their problem was research and knowledge, not design. This is how students also saw research. If researchers were not venturing into obscurities in research, students appreciated their work and saw value in it. *Prototyping Interactions* and *Design Research through Practice* were intelligible to designers, and both helped to expand their reach. In this environment, building design research on making things would have been nostalgic and regressive. Even today, the empathic group continues to give people control and sees prototyping in a human context. By now, however, it also knows how to turn prototyping into a vehicle of interpretive knowledge.

Smart Products

Usability of Smart Products

Maypole

One-Dimensional
Usability

The Future of
Digital Imaging

Cardboard Mock-ups

eDesign

Design Games

IKE

Empathic Design

Design Studio
in the Field

Luotain

Mobile Image

Co-experience

Design Probes

Collaborative
Design

Radiolinja

Prototyping
Social Action

Women
and Jewelry

You are Important!

Design for Home

Dwelling with Design

Pasadena

Morphome

Electrome

IP08

Art of
Research

Prototyping
Interactions

SRILE Project

Design Research
through Practice

Bikes and Plants

Against Method

Memories in Clay

Drifting by Intention

Folk Tradition

Runkokuituja

What Happened

Novapro

From Disposable
to Sustainable

DWoC

Lost in Woods

6 Interpretive Design Research and Beyond

This final chapter shifts the focus from the empathic group to the main theme of this book, the interpretive framework. It concludes the book by examining three questions.

The first question is the framework behind an interpretive design research program. This book argues that an interpretive design research program must be user-centered, have research techniques to study humans, be open to collaboration, and have ways to integrate making things into its foundation.

The second question studies the program in the context of larger shifts in design research. Over the last decade, design research has moved from being user-centered to systemic, and there are calls to turn design into a discipline that serves nonhumans. This chapter shows how interpretive researchers can respond to these challenges.

The third question studies the relevance of an interpretive approach to design research and makes a plea for its continuing relevance. The final section argues that an interpretive approach provides a way for keeping dehumanizing and potentially totalitarian patterns of thought at bay in design research.

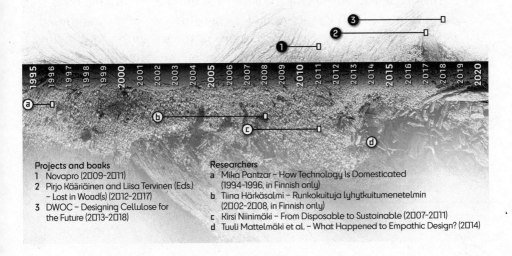

Projects and books
1 Novapro (2009-2011)
2 Pirjo Kääriäinen and Liisa Tervinen (Eds.) – Lost in Wood(s) (2012-2017)
3 DWOC – Designing Cellulose for the Future (2013-2018)

Researchers
a Mika Pantzar – How Technology Is Domesticated (1994-1996, in Finnish only)
b Tiina Härkäsalmi – Runkokuituja lyhytkuitumenetelmin (2002-2008, in Finnish only)
c Kirsi Niinimäki – From Disposable to Sustainable (2007-2011)
d Tuuli Mattelmäki et al. – What Happened to Empathic Design? (2014)

It is time to return to the main question of this book. What would design research look like if it were built on an interpretive foundation?

Answering this question requires stepping out of the case that this book has been analyzing. Up to now, it has been weaving the story of empathic design with a larger change in design research. This may have given the impression that empathic design was reacting to and following larger trends. The first two chapters of the book described how the group developed its agenda in the context of information technology and the emerging concept of user experience. The next three chapters described its adoption of codesign processes, design-based research methods, and a constructive, prototyping-based approach to research. All these developments were described as a part of larger, global changes in design research.

The group has been more than a follower, however: it has shaped design research. It has made several significant contributions that have shaped design research far beyond its boundaries. Many of its contributions have become part of the standard repertoire of design research globally, including the mobile multimedia studies that I did with Esko Kurvinen and Turo-Kimmo Lehtonen (Koskinen, Kurvinen, and Lehtonen 2002; Kurvinen 2007; Koskinen 2007); Katja Battarbee's concept of co-experience (Battarbee 2004); Tuuli Mattelmäki's interpretive perspective on cultural probes (Mattelmäki 2006); the book *Empathic Design* (Koskinen, Battarbee, and Mattelmäki 2003); and methodological research that I did with Thomas Binder, Johan Redström, John Zimmerman, and Stephan Wensveen (Koskinen et al. 2011). In terms of societal impact, Katja Soini's (2015) *IKE* project is the leading candidate.

These contributions, however, do not address the question that this book has tried to answer. The main reason for telling the story of the empathic group goes beyond its experience. It is one of the few groups in design research that has consistently built on an interpretive theoretical foundation, which has shaped its research in numerous ways. The group gives us a unique opportunity to develop an interpretive framework for design research and to assess its value in the broader panorama of design research as it is currently practiced.

This chapter first describes an interpretive framework that the program has spawned. Then it probes the value of the framework through two questions: first, what is its value in the context of contemporary user-centered design;

and second, what is its value in the shifting landscape of design research at large: is this framework still valuable after more than twenty years, during which design research has shifted away from its user-centered basis? After examining evidence, the chapter makes the case for the continuing relevance for a human, interpretive framework like the one presented in this book.

Four Sensitivities: An Interpretive Design Framework

One answer to how interpretive design research would look is the framework at the end of "What Happened to Empathic Design?" a paper I published with Tuuli Mattelmäki and Kirsikka Vaajakallio in *Design Issues* in 2014. Toward the end of the paper, we described four sensitivities that have given shape to the program. Taken together, these sensitivities provide a framework that tells what any interpretive research program should pay attention to. They are sensitivities toward humans, technique, collaboration, and design. Figure 6.1 describes these sensitivities, shows their connection to research practice, and links them to the content of this book.

The drawing conveys solutions to a few problems that any interpretive research program will face. The first element is an interpretive theory that helps researchers keep track of dozens of theoretical models in contemporary design research. In this book, theory has been coming from symbolic interactionism and Blumer (1969), but there are other options. This book has mentioned symbolic anthropology at Intel, Bruner's narrative psychology in *Empathic Design* (2003), Csikszentmihalyi and Rochberg-Halton's (1981) framework for studying the meaning of objects at home, and, of course, ethnomethodology (Kurvinen 2007). Symbolic interactionism is only one potential interpretive foundation, but the way that it has guided the empathic group over the years shows its importance. It provides consistency of vision, helps to specify complex concepts, and keeps research coherent over the years. Without familiarity with theory, it is easy to get lost in the endlessly rich world of human experience.

The second element is a choice regarding how to approach methods that support the interpretive theory and turn it into concrete research operations. There is no shortage of research methods, most of them having their origin in the social sciences, and design researchers need find a way that is conducive to design. This book corroborates one of the key lessons of participatory

Figure 6.1

HOW ABOUT **SENSITIVITY** TO **COLLABORATION?**

INTERPRETIVE DESIGN RESEARCHERS COLLABORATE AND COMMUNICATE WITH PEOPLE TO ENSURE THEY UNDERSTAND THEM. THEY USE **DIALOGICAL, CODESIGN, COLLABORATIVE AND CONVERSATIONAL DESIGN METHODS** TO ACHIEVE THIS.

WHAT DO YOU MEAN BY **DESIGN SENSITIVITY?**

THOSE COMPONENTS OF THE INTERPRETIVE FRAMEWORK THAT GO BEYOND ITS **THEORETICAL BASIS:** TECHNIQUES, COLLABORATIVE APPROACH, AND CONSTRUCTIVE COMPONENTS. THEORETICAL BASIS HELPS IN UNDERSTANDING HUMANS, BUT **DESIGN RESEARCH** ALSO NEEDS WAYS TO LAND KNOWLEDGE OF HUMANS INTO **DESIGN PRACTICE.**

Interpretive Design Research: A Framework

CAN YOU GIVE **EXAMPLES?**

CHAPTER 3 DESCRIBES THE PATH FROM ACTION RESEARCH TO COLLABORATIVE DESIGN. SOME EXAMPLES ARE TUULI MATTELMÄKI'S **DESIGN PROBES** (2006), ANDREA JUDICE'S **DESIGN FOR HOPE** (2014), AND KATJA SOINI'S STUDY OF HOW COLLABORATION WORKS IN A NETWORK (2015).

THERE ARE LOTS OF TOOLS, INCLUDING EARLY-STAGE, COLLABORATIVE AND CREATIVE METHODS IN **CHAPTERS 2-4.** TRADITIONAL DESIGN TOOLS LIKE SKETCHING, BUILDING MOCK-UPS AND PROTOTYPES CAN BE SEEN AS VEHICLES FOR CREATING KNOWLEDGE, AS **CHAPTER 5** SHOWS. THINGS TO READ ARE **CARDBOARD MOCK-UPS and CONVERSATIONS** BY SIMO SÄDE (2001) AND **DESIGN RESEARCH THROUGH PRACTICE** (KOSKINEN ET AL. 2011).

WHAT SHOULD **I READ** FIRST?

designers, who tell designers to use sketches, mock-ups, and low-fidelity prototypes when interacting with people and leave theories to the university (Ehn 1988; Ehn and Kyng 1991). These methods have included low-fidelity prototypes like cardboard mock-ups and paper prototypes, experience prototypes, narrative techniques, and virtual platforms for studying social actions (Koskinen et al. 2003). With these methods, the researchers have stepped down from their ivory towers to a level at which they can talk to people not from above, but as equals. This sensitivity is similar to how anthropologists describe their preference of emic concepts over etic—that is, concepts that people use, as opposed to concepts from the research world (cf. Pike 1954).

The third element is collaboration. Reports, presentations, computer-aided design (CAD) models, videos, and prototypes convey design outcomes, but they do not address the gap in understanding between the designers and audience (e.g., customers and other researchers). Interpretive design researchers need to find a way to address this gap to make their ideas effective. The precedent that the empathic group has found useful has been codesign. By designing along with the audience, the audience will learn the rationale behind the design ideas, so they do not have to reconstruct it from the outcomes.

The final element is the researchers' relationship to design. An interpretive approach to design research rejects the idea that designers are technical experts whose source of authority is their superior knowledge, as in the traditional professions. Its alternative is to see users and other stakeholders as a source of authority, and designers as interpreters. As interpreters, their claim to authority lies in their ability to bring people and the material world closer together. The measure of success is whether they can show people the way to live their lives in more meaningful ways. The source of meaningful change is people themselves. A good translation of a book opens another language and culture to a reader who would not have access to them without the translation; a good piece of design research opens up material and digital worlds in the same way. The skill of a translator lies in the ability to produce versions of works that open up new worlds. This self-image subjects technical expertise to the wider goal of advancing human understanding and is in fact more demanding than the traditional role of designers as problem solvers.

It is important to note that the framework proposed in the 2014 paper by Mattelmäki, Vaajakallio, and me has its origin in Herbert Blumer's sociology, but it goes beyond it. Blumer gave the empathic group tools that allowed them to be sensitive to humans. Yet other elements of the framework are relevant to designers rather than sociologists. For example, elements like sensitivity to technique and sensitivity to collaboration take a unique form in design. In the framework, techniques are valid if they help to translate the human world into design insights. These insights become the measure of success of the framework, and its value depends on its contribution to design research. Sociological theory can provide a foundation, but it is not enough. As a discipline, design requires projective and constructive elements that push sociological insights into concepts, products, processes, and other types of design outcomes. This is what the three additional components of the framework do: they turn a sociological approach into a design-specific framework.

The Growth of Interpretive Knowledge

Research creates knowledge, and the main measure of any research program is whether it creates new knowledge. How can we measure progress in interpretive design research?

Contemporary design research has several ideas about how it creates knowledge and how knowledge grows. As I have argued with Peter Krogh (Krogh and Koskinen 2020), design researchers see knowledge in many ways. At one end are researchers who believe that knowledge is carried by the actual designs. At the other are those who locate knowledge in frameworks—the four sensitivities framework discussed earlier in this chapter is an example—and debate these frameworks. Still others see knowledge as a property of the communities they are studying. A few researchers have even dreamed of turning design into a mathematical discipline (Overbeeke, Wensveen, and Hummels 2006). In this context, interpretive design research has a dialectic concept of knowledge. Its goal is to create design frameworks that make sense of people and their experience rather than acting like legislators who can impose their viewpoint over these people because they have authoritative and superior knowledge. Previous studies provide useful cues, but they function as references rather than axioms.

Yet, as this book shows, the empathic group has expanded its reach significantly over the years. There is an internal logic in its expansion. Can the expansion be described as a "cumulation of facts," as in mathematics or the natural sciences? In the minds of the group members, the answer has been no: the progress has not been cumulative in the same sense as in the natural sciences, which progress by solving problems and treating solutions as if they were Lego blocks that can be reused anywhere. The group found a better way to describe its idea of progress from interpretive anthropology. For Clifford Geertz (1973, 25), anthropological knowledge does not grow by accumulation like in science. Instead, it grows

> in spurts. Rather than following a rising curve of cumulative findings, cultural analysis breaks up into a disconnected yet coherent sequence of bolder and bolder sorts. Studies do build on other studies . . . in the sense that, better informed and better conceptualized, they plunge more deeply into the same things . . . the movement is not from already proven theorems to newly proven ones . . . a study is an advance if it is more incisive—whatever that may mean—than those that preceded it; but it less stands on their shoulders than . . . runs by their side.

It can be argued that this view is familiar to designers as well. For example, industrial design has seen several growth spurts since the 1950s. Industrial designers have advanced their practice by studying information theory, semiotics, ergonomics, management, computer-aided design/computer-aided manufacturing (CAD/CAM), user-centered design, emotional psychology, surrealist methods, action research, service design, and sustainable design, to name a few.

During its twenty-plus years, the program has generated a significant body of knowledge. Using that, it can answer new types of questions by using previous studies as references and occasionally as precedents. When it has not found answers in the body of knowledge that it possessed, it has expanded its scope. While in 2000, the nascent program had to develop methods for studying user experience of smart products, in 2020 it had a record of accomplishment in the areas of services, design processes, co-experience, service design, social design, collaborative design, world design, several types of prototyping, and even conversational and collaborative art. The history of empathic design does not prove that other interpretive approaches to design research will advance, but it does suggest that advancement is possible. This advancement is one of the reasons for telling the story of empathic design. An interpretive approach has a legitimate place in the larger gamut of design research if the criterion is progress.

The history of the empathic group shows that a genuinely interpretive research program in design can indeed create new knowledge. The program has made significant contributions to design research, and it has done much more than just apply standard design methods to new problems. It has shifted its problems several times in response to changes in its environment or internal logic, and its interpretive framework and its main lines of research are much more incisive—to use Geertz's term—today than they were twenty years ago. Because of this, they can respond to problems that would have been out of reach only ten years ago, and they use a much wider array of tools. Of course, the success of the empathic program cannot prove that other interpretive programs would progress in the same way, but it shows that progress is possible.

Interpretation after Twenty Years

What kind of contribution have the group and its framework made to user-centered design research? The interpretive foundations of empathic design are about twenty years old. These foundations have solved several problems, but they have also uncovered other problems. The empathic group's interpretive approach to design research was fresh in 2000. Back then, there were few coherent interpretive approaches in design schools, but design research has matured since then. Is the program still novel and necessary, or have other developments caught up with it?

The best place to start answering this question is a scene from the beginning. In 2000, ecological, emotional, and hedonic theories had broken the dominance of cognitive psychology in design research (Dandavate, Sanders, and Stuart 1996; Norman 1998; Jordan 2000; Overbeeke 2007). Companies like Nokia and Philips already had been advocating psychological approaches to research for a few years. The best work in distributed cognition and social action revealed at the Computer-Supported Collaborative Work conference was sophisticated, but these approaches had found little following in design schools, perhaps because it took research training to understand them (Suchman 1987; Hutchins 1996). The empathic group was following several companies in Chicago and Silicon Valley, Karen Holtzblatt's contextual inquiry, and activity theory (cf. Kuutti 1996), and it knew about the Canadian professor Jorge Frascara's *Design and the Social Sciences: Making Connections* (2002). An openly interpretive sociological approach was novel in this context.

A few years later, the situation was starting to change. Design anthropology was maturing rapidly and convened for the first time at the Ethnographic Praxis in Industry conference in Redmond, Washington, in 2005.

The craft research community found its voice at the Art of Research Conference in 2005 and in the *Craft Research* journal, which started in 2010. John Dewey's pragmatism found proponents in design-oriented human computer interaction (HCI) (Forlizzi and Ford 2000; McCarthy and Wright 2004; Battarbee 2004; Dalsgaard 2009; Dixon 2020). Several authors also introduced phenomenology and postphenomenology to design (Gedenryd 1998; Lenay 2010). All three strands have been strong in neighboring disciplines like craft research and HCI (see, for example, Mäkelä 2003; Winograd and Flores 1987). Design research is theoretically far richer today than twenty years ago.

In contrast to the year 2000, when design research was dominated by user-centered approaches and usability testing, today it is normal to see design as a research instrument. This is true not only at the artistic and craft-oriented end of design research, but also in HCI. There is a lively debate about the methodology and even the epistemology of design as a method. Similarly, codesign has gone mainstream by this point. A contemporary approach to design research brings together knowledge of humans, invites stakeholders to participate in the research process, and uses design as a method with sophistication. Empathic design was one of the programs that helped to define this approach. The four sensitivities of Mattelmäki and her colleagues are not unique individually, but their combination is.

Still, even today there are gaps in research. One of the obvious ones is understanding society as a design opportunity. I gave a talk about social design a few years ago to a leading interaction design department in the Netherlands. There were sixty to seventy people in the audience. After the talk, the discussion turned back to the meaning of "social." After a couple of questions, I asked the people in the audience how many of them had studied sociology. No hands. Political science? No hands. Economics? One hand. The department was acutely aware of the importance of social processes, and it was building part of its research program around it, but where were the tools?

A year after, I posed the same question to a group of doctoral students in another leading design department in Italy. The result was the same. Over beers after the meeting, my former colleague Alastair Fuad-Luke wryly noted to me that I had pointed out an elephant in the room. (As a former plant ecologist, he had faced the same discussion in the context of biology.) Perhaps this is where the shift from empathic to interpretive design has made its most significant contribution. A coherent approach for studying people in social processes is still missing, by and large, from design research.

What conclusion can we draw from this? The program described in this book, which puts design research into a social context, remains important. If we do not have theoretically elaborated frameworks for studying humans, we fall back on our lay notions of people and society. As I noted in an introduction to a special issue on social design for the *International Journal of Design* in 2016 with my Taiwanese and Dutch colleagues, understanding modern society requires knowledge of the social sciences:

> [A]ctions are parts of long chains of action and rules of governance that make it difficult to see the consequences of the actions. Pressing a button may move billions in a millisecond, and while it creates profits for the owners of the mining company, it may also slash three thousand jobs in a mining town in the Andes . . . Two grand modern structures of governance, the state and the market, often stand beside community networks and enter into the very constitution of social problems. Understanding these kinds of complicated linkages is the bread and butter of the social sciences, but designers are still ill equipped to deal with them. (Chen et al. 2016, 3)

As we have pointed out, if designers want to study the complexities of modern structures of governance, they need a knowledge of society that does not usually exist in the design world. In the long run, design researchers cannot continue doing projects on a small scale while promising large changes in society. Rather, they need to find ways to tackle large-scale issues in society and find ways to contribute to processes that are capable of making these changes. The best way may be surprisingly close at hand. Looking at the history of the empathic group, Soini (2015)'s study of resident-centered housing renovations provides an important cue about how to leave the world of small projects by coopting political and industrial decision-makers. Another recent interpretive study that has been successful in crossing the bridge from the scale of a village to modern structures of governance proposed to conduct codesign on the massive scale of the European Union (Meroni, Selloni, and Rossi 2018), while a third study built a design-led think tank for the Danish government (Bason 2017).

If we follow these three researchers, the key is to shift our attention from users to parties and processes that make and change the rules that constitute society. We need to find ways to coopt them. These are new frontiers to design researchers, but researchers like Soini (2015), Meroni, Selloni and Rossi (2017), and Bason (2017) have shown that an interpretive approach can respond to issues that are much more complex than the problems that researchers were dealing with in 2000.

How Interpretation Can Meet Recent Challenges

In chapter 1, I noted that design research has been moving to new topics since the end of the 2000s. The main concern of design researchers in 2000 was information technology and its impact on their work, but by 2010, the concerns were shifting elsewhere. Sustainability was becoming a major concern to many in the aftermath of service design, identity politics was finding its way into design, and many design researchers started to take ethics more seriously, among other developments. It is hard to pinpoint any one reason for these changes, but a few potential candidates stand out. On the human side, the 2010s has seen a proliferation of questions about identity, which have led to questions about potential biases in design. Scientific consensus about climate change had existed for a decade, and while it has taken time to trickle down to design, it has become a major concern in recent years. It has also led to even wider questions about the planet and the goal of design: is the goal serving human beings, or should the goal be serving the planet in a way that sustains life?

These forces started to turn to new research topics in a few years, but none of them was directly related to the main driver of the interpretive framework described in this book. The main concern of the empathic group was user experience—how to create products that are pleasant and exciting to use and that help people gain a competitive edge in the marketplace. In this world, there was little mention of sustainability, inequality, or questions about identity: who are we, and how do we see ourselves? It could even be argued that user-centered design behind the empathic program ran counter to many of these trends. After all, it was devised to help industry create more attractive products to be sold to a generic user without considering qualities like age, gender, or ethnic identity.

These debates have exposed potential shadowy regions in the interpretive framework outlined in this chapter. Still, there may be ways to turn these challenges into opportunities. The following section examines the empathic program to learn what these opportunities could be.

Users and Identities

Interpretive design research has its origin in user-centered design. It has added a considerable amount of knowledge about how people use technology in many circumstances of their lives, ranging from using modern telecommunications to healing from tuberculosis. But the very word "user" demands clarification. Who is the user? Does the user have gender, age, history, sexual orientation, or ethnicity? Who is the user that design researchers want to make sense of, and if we only study white, middle-class people in First-World countries, does this bias the way in which we construe users (Redström 2006)? Are we unwittingly turning a small portion of humanity into a yardstick that does not serve humanity at large? Shouldn't we instead expand design to serve people who do not have access to it?

The interpretive framework outlined in this chapter can be used to study many kinds of people. Research practice, however, is another matter. How has the empathic group, with its base in the middle class of a First-World city, expanded its notion of the user, and what does its experience say about how interpretive researchers can respond to the criticism that their human basis is skewed?

One potential source of bias is occupational, but except for the few first years, the group was always multidisciplinary. It originally consisted of industrial designers who worked with interactive technologies and smart products, but it soon included sociologists and constantly interacted with psychologists and economists, as well as computer scientists, mechanical engineers, electronic engineers, and microbiologists. Also, the context in which the group was working, a major design school, brought it into contact with several other design disciplines, and the line was permeable. In the early years of the program, the group partnered mostly with craft designers, multimedia designers, graphic designers, artists, and actors. These collaborations introduced the group to the crafts that have traditionally been driven by artistic sensitivities and manual skills, rather than an interest in users or humans in general. Yet most of these people were of European origin until about 2004, when Korean, Brazilian, and Columbian researchers joined the group, and later, researchers from India, Canada, and Norway, among other nations, were added as well.

In terms of participants in its studies, the group has been more expansive. Between 1999 and 2002, participants in studies like *Maypole, eDesign,* and *Mobile Image* came from several parts of Europe, but they were invariably white. Between 2004–2008, the base expanded to the United States and to the ethnic enclaves of Helsinki, Brazil, Estonia, and Denmark. At the end of the 2000s, the group studied women and the middle classes in India. The methodological studies after 2007 were built on a highly diverse student body that was half European, half Asian. Tangential to the empathic program were researchers who conducted studies in Namibia and Iran (Miettinen 2007; Sorainen 2006).

A few studies have targeted identity directly. Maarit Mäkelä's *Memories in Clay* (2003) studied the matrilineal line of her own family to understand female experience, Inkeri Huhtamaa's *Namibian Bodily Appearance and Handmade Objects* (2010) investigated the relationship of craft and bodily adornments in Namibia, and Kärt Summatavet's *Folk Tradition and Artistic Inspiration* (2005) examined gender symbolism in traditional Estonial jewelry. She showed how symbols, materials, and techniques reflected Estonian folk beliefs about a woman's life. In these studies, craft and feminism came seamlessly together. These studies also brought about a crucial difference in the group's user-centered methods. The group validated its findings with people: its source of truth was external to the design world. For a craft researcher like Summatavet, the criteria of success were the opinions of her peers. Yet learning went both ways with craft researchers. The group saw how the studio was used as a tool in craft research. Craft researchers in turn got methods, research models, and international networks from their empathic colleagues. The locus of proof was different, but the relationship was symbiotic.

A few studies from the industrial designers in the group expanded the user base conceptually. Jung-Joo Lee's doctoral work (2012) was a sustained effort to understand whether Western design methods can be adapted to an Asian context, and Marcelo and Andrea Júdice studied the impoverished village of Vila Rosario to improve its tuberculosis treatment practices (M. Júdice 2014; A. Júdice 2014). Their studies raised questions about the social basis of ethnicity in Brazil and whether it should be a duty of designers to serve people outside their usual circle of recipients. Other work from the fringes of the group expanded the basis of design to incorporate design for all. Helena Leppänen developed crockery for senior citizens (2006), and Susanne Jacobson (2014) borrowed the concept of stigma from the sociologist Erving Goffman (1963) to study how to design objects that do not stigmatize disabled users.

These studies leave several gaps. For example, students were interested in sexual identities and sexual violence, but these topics were difficult to address in the functionally oriented context of user-centered design. Still, as the examples I have mentioned here show, an interpretive approach can expand relatively easily into questions of identity. Identities and social identities are like any other objects: as soon as people start to think, talk, and write about them, turn them into images and objects, and place these constructs into broader social contexts like rituals and religious beliefs (Priha 1991), these identities start to take forms that can be approached only by explicating them. People make sense of themselves and their relationship to these construed selves. As researchers, we can interpret these meanings, as well as their consequences and repercussions. If we shift research from product use to identity as a determinant of design, we may lose the simplicity of the concept of "user" but win an immensely rich view of human beings.

Sustainability, Systems, and Science

Another pitfall of user-centered design is that it usually focuses on people in the context of using technology. For example, an ethnomethodologist like Lucy Suchman (1987) would study how people make sense of a copying machine while using it, and how these situated actions rather than knowledge explain whether these people get their copies from the machine or not. User-centered approaches have expanded the scope of design considerably over the past three decades. Again, however, the problem lies in the notion of "user." Many problems that designers face today are not solvable by focusing on the user. Rather, they require systemic solutions that cannot be studied through traditional, user-centered tools. What if user experience is an end-of-the-pipe phenomenon with its origins in systemic properties that are outside the control of humans who experience things? This criticism is familiar from the social sciences, where Marxists have often dismissed interactionism as being incapable of targeting systems even though they are more important for understanding society than those use situations that people like Suchman were interested in.

The group had doubts about system analysis that tends to find order in places where it does not exist. When applied to society, it is famously conservative and tends to see factors that push systems out of equilibrium as pathologies (Gouldner 1970). With this caveat, there is one place in which the empathic program tackles the systemic challenge: sustainable design. Design research started to shift to sustainability and services in the 2010s. Its dominant mode of research was systemic rather than user-centered. Its main concern was not whether we can create better technologies to improve user experience; rather, it was pushing the design community toward saving us from the excesses of technology for environmental purposes. For example, the ethos behind service design was to diminish the human footprint by creating services rather than material things (Vezzoli et al. 2014; Manzini 2015; Lettenmeier 2018).

The group addressed the sustainable challenge through three lines of research. The first put objects in the context of emotional ecology. A direct outgrowth of the group's empathic approach, it explored emotional connections to objects as a source of dematerialization (see also Chapman 2005, 2021). Katja Battarbee explored stories as a source of emotional data in her paper "Stories as Shortcuts to Meaning" (2003b). One of the cases she explored was a rocking chair that had been kept in her friend's family for over 100 years and was still being used, and her friend told it was the history of the object that made it meaningful, even though it would have been easy to buy a new chair. Heidi Grönman and I studied this idea later in a paper that explored how design designations may play a role in adding longevity to objects (Koskinen and Grönman 2006). The doctoral theses of Paavilainen (2014) and Ahde-Deal (2013) were in part an outgrowth of this hypothesis, which was also explored by Kirsi Niinimäki (2011) from an individual angle.

The second line was design activism. This idea of activism partly came from Alastair Fuad-Luke (2009), who had joined the design school of Aalto University in 2011, but also from two doctoral theses by Miettinen (2007) and Sorainen (2006), which had built on activist principles in developing tourism as a livelihood in Namibia and in salvaging pottery as a livelihood in Kalporagan, Iran, where a 4,000-year tradition was threatened by cheap imports. The first theses that came from this work were by a new group of students. Garduño García (2017) developed an ethical framework that she tested in a poor and marginalized Mayan community in Campeche, located at the root of the Yucatan Peninsula in Mexico. Kohtala (2016) studied how Fab Labs, a network of digital fabrication laboratories that originated at the Massachusetts Institute of Technology (MIT), negotiated the gap between their sustainable ideology and technological reality. Acharya (2017)'s thesis added an artistic approach to activism.

The leader of the activist line became Kirsi Niinimäki, who joined Aalto University's design school as a professor in 2015. She gathered a group of doctoral students who expanded design activism to textile and fashion design. Aakko (2016) studied craftivism as an alternative to a resource-depleting fashion system based on the current system. Durrani (2019) studied the mending and repair practices of clothes among craft activists in Helsinki, New Zealand, and Edinburgh, and Valle-Noronha (2019) created wardrobe interventions in Helsinki and Belo Horizonte, Brazil, to forge an active engagement with clothing. Hirscher (2020) explored how to integrate refugees to society in fashion workshops in Helsinki, Germany, and Italy, and Chun (2018) studied place as a source of value in the local production of clothes in Helsinki. Activist ideas were significant in this research not in themselves, but because they led to design objects and worked with social conditions that either enabled or hindered them.

The experience of the empathic group has thirdly involved scientific knowledge that has expanded its research from human beings into nature. Part of the recent development of the group has brought it into contact with chemists, material scientists, physicists, and microbiologists. Still, such collaborations were not completely new to the group. A group of ceramic designers that it was connected to had explored turning recycled construction materials into bricks and tiles, atomic-level glazing methods in the kiln, and ways to recycle glass waste from monitors by turning them into toxic waste containers (Sevelius 1997; Eerola 1999). After receiving their doctorates, these designers became teachers or took jobs in industry, and their work in sustainable design found its culmination in Raija Siikamäki's thesis *Glass Can Be Recycled Forever* (2006). These designers had been mentored by Markku Rajala, a material physicist who joined the ceramics program in 1989, but he left in 1997. After his departure, the students went without support, and the line could not attract new students.

The legacy was not wasted, however. A small group of textile designers launched a series of collaborations with natural scientists in which interpretation plays no role. The initial goal of these collaborations was to find ways to replace cotton in textiles with fibers made from cellulose to reduce the environmental impact of cotton. Empathic design played a role in the initial stages of this collaboration, but it took a scientific and speculative direction later (for more information, see Kääriäinen and Tervinen 2017).

The pivotal role in these collaborations fell to Tiina Härkäsalmi, a textile designer who developed methods for producing bast fibers from flax and hemp. She started experimenting with fungus and got it working in a small laboratory in her garage. She published her findings in *Runkokuituja lyhytkuitumenetelmin* (2008, Bast fibers by short-fiber methods), and set up a collaboration with microbiologists and material scientists in the *Novapro* project. In this project, the scientists did not find the causal mechanism behind the fungus's ability to break fibers, but it was valuable in two other ways. It produced enough fibers for weaving, knitting, and coloring experiments, and it showed that the program's interpretive basis can support scientific research. The progress was not always smooth, but Härkäsalmi found a few natural design roles in this collaboration. Most important, she helped the scientists keep people in mind. Other fruitful roles were finding potential applications for scientific knowledge, designing pieces that helped to popularize the scientific findings, and opening up new sources of inspiration for scientists.

Härkäsalmi's experience paved the way to other science collaborations led by Professor Pirjo Kääriäinen, a textile designer and design manager. One was the *DWoC* project, which brought chemical engineering and design together to create design-driven material research (Kataja and Kääriäinen 2018). It also developed sustainable uses for cellulose, the most abundant organic polymer on Earth. In addition, it introduced experimental prototyping to speed up material development and to probe business models and market opportunities to new, cellulose-based materials generated through scientific methods that included vapor-driven curling of nanocellulose, superhydrophobic surfaces, and novel machinery for product manufacturing (*DWoC* 2016). The European Union's *Trash-2-Cash* project (https://www.trash2cashproject.eu/) produced and used regenerated fibers from textile waste. These projects also led to the CHEMARTS program, which institutionalized the collaboration (Kääriäinen and Tervinen 2017; Kääriäinen et al. 2020) and led to the establishment of a bio-innovation center and to an interdisciplinary doctoral school (Aalto University 2020).

Collaboration with scientists has shown the empathic group that its frameworks and methods had some limits. As a rule, the group has gone back to its roots, and it has played an auxiliary role in these collaborations. Interpretive user research and design experimentation have clearly been valuable for scientists. The interpretive approach of designers has helped them to focus on societal problems, and design experimentation has given them a quick feedback mechanism about the experiential qualities of their work. Designers in their turn have learned a new way to look at time. Scientists have taught them patience in order to keep their eyes on long-term goals.

Interpreting Nonhumans

A third recent debate that has exposed a potential gap in interpretive design research may be the farthest reaching. A substantial part of the most exciting new design research has argued that designers need to encourage society to live in harmony with nature by giving a voice to nonhuman entities. These cover organic and inorganic nature and human-made artificial things that inhabit our life-worlds (Colomina and Wigley 2016; Forlano 2017; Giaccardi and Redström 2020; Wakkary 2021). At its base, the critique says that user-centered design has contributed to many of the ills that are ruining the planet, and it is time to find ways to design with nonhumans like animals, rivers, and systems on a geological and regional scale, and give these things a status equal to humans (Thackara 2019).

The most forceful recent spokesperson for taking nonhumans seriously, as cohabitants, is the Canadian design professor Ron Wakkary. In his recent book *Things We Could Design* (2021), he notes that designers have privileged human values by promoting human-centered approaches into design. There is an elephant in the room, however. Human-centered design is threatening those very values that it is promoting by contributing to a worldview that puts humans over nature. The last five decades have shown that humanity is depleting the nonhuman and natural world by subsuming it to our own needs. Many species have become extinct, and although wealth and technological prowess may shield some societies for a while, climate change is threatening everyone. If Wakkary is right, design must change its course radically and learn to design with nonhumans in mind.

It goes without saying that unless we are careful, interpretive design research fares badly in the context of this argument: it starts from human activities and focuses on them. Is it old-fashioned and does its human ethos work against its larger goal of creating a world that is better to live in? Is it part of systems that create problems? If so, how could it become part of systems that create solutions? Of course, it is impossible to answer these questions directly because the empathic program predates this critique by twenty years; however, by examining its history more closely, we can see at least two routes that might provide an answer that Wakkary could accept.

One of these has been indirect. Nature and nonhumans do not speak to each other. People have vocal cords, books, elections, representatives, and social media sites, but nonhumans as a rule need someone to give them a voice. The work of Tiina Härkäsalmi, Kirsi Niinimäki, and Pirjo Kääriäinen is especially relevant. Another project that collaborated with scientists was *Menomaps* (2008–2009), in which Salu Ylirisku developed interactive maps for a national park for geologists to use. These researchers have shown that design researchers can give voice to nonhumans by working with scientists, activists, and other groups that have a deep knowledge of nature and tools to make its processes tangible. They do so by participating in the worlds of groups that have knowledge of nature, whether organic or inorganic, and that can work at a molecular and even subatomic scale, as well as with units the size of a national park. Any interpretive design researcher can follow this example and develop tools to create nourishing habitats for endangered species, protect wetlands needed to protect human settlements, clean toxins from lakes and plastic from seas, and other positive actions. There are also possible theoretical pathways, but these have been less productive. Bruno Latour (1987)'s actor-network theory is symmetrical, in that it gives non-humans and humans similar powers of agency (think about a speed bump on the road). This theory was discussed in doctoral seminars, as well as by authors like Donna Haraway (1988), but although the group found it easy to understand conceptually, it was too leadership-centered and Machiavellian for the group's taste, and it was not fine-grained enough for design purposes.

This approach is indirect, but it has an important precedent in design. In the 1960s and 1970s, participatory designers worked with unions to give voice to blue-collar workers in the development of technology (Ehn 1988). Designers as such have little power, but they can work with people who do have power. Perhaps the way forward is to work with environmental activists, scientists, and proenvironmental politicians to give voice to the nonhuman world. It is also good to keep in mind that while the power of the unions has been decreasing almost everywhere, the power of people who give voice to nonhumans has been increasing in most countries. Rachel Carson's groundbreaking book *Silent Spring*, the Sierra Club, and the Club of Rome put out important wake-up calls about the need to preserve nature, but these were lonely voices. Today, however, concerns about the climate have become mainstream. It is not difficult to find experts with intimate knowledge of beach protection levies or the intricacies of greenhouse gases in the atmosphere.

Another route has been theoretical. It can again be illustrated with a piece of history from the empathic program. *Kuinka teknologia kesytetään* (How technology is domesticated) was a book by the economist Mika Pantzar, who worked as professor of industrial design in 1997–1998. He was interested in how radical innovations become practices that reshape society. The world is not short of inventions, nor is it short of inventions that are claimed to be radical, but some of these inventions do become second nature to us. It would be hard to imagine a modern world without cars, airplanes, the internet, television, cellular phones, or social media. Pantzar's analogies were the domestication process of domestic animals and the way in which forests regrow after fires (Nieminen-Sundell and Pantzar 2003). He knew well enough that creating an explanation of how technology is domesticated goes far beyond the ken of even leading philosophers. Instead, he pooled together a formidable collection of papers and books and wrote a book-length essay to discuss what the world's leading researchers have said about why some technologies stay and some disappear.

The answer to *Kuinka teknologia kesytetään* came from many sources, including economic literature on innovation, Roger Silverstone's (Silverstone and Hirsch 1992) domestication studies of media technology, sustainable consumption literature, and a host of new perspectives in what can loosely be called social studies of technology and social studies of technology and science. The book also referred to Mihaly Csikszentmihalyi on the meanings of things, Arjun Appadurai's anthropology of objects, Igor Kopytoff on the biography of things, Bruno Latour's and John Law's versions of actor-network theory, evolutionary economics, Donna Haraway on biological interpretation of humans, and a host of sources from historical and sociological studies of science and technology (Csikszentmihalyi and Rochberg-Halton 1981; Appadurai 1999; Kopytoff 1986; Latour 1987, 1990; Callon 1987; Boulding 1989; Pantzar 1991, 1993; Lynch 1993; Haraway 1988; Bijker, Hughes, and Pinch 1987).

These studies all told pointed to the same conclusion: if we want to understand domestication, we need to study what consumers and ordinary people do with technology and turn away from focusing on industry and commerce.

Pantzar's book has not been translated into English, but its main ideas are well known in the global design research community through its influence on practice theory. The influence has its origins in Pantzar's collaboration with the British sociologist Elizabeth Shove. In *Manufacturing Leisure* (2005), a book that Pantzar and Shove edited, they explored several ways in which leisure had been turned into products and services. The book developed an empirical way to study questions that Pantzar had raised earlier. Shove led a series of meetings in Durham, England, that inspired her subsequent work with geographers and designers (e.g., Shove et al. 2007). The collaboration between these two researchers led to *The Dynamics of Social Practice* (Shove, Pantzar, and Watson 2012), which gained a worldwide following in the 2010s. The empathic group knew practice theory well, but it remained on its fringe (however, see Marttila 2018). The main reasons were probably timing and the nature of the unit of analysis behind the theory.

However, Pantzar's book opened theoretical vistas that went beyond user-centered design and the social sciences behind them. For example, he was skeptical about *homo economicus*, and he warned about technological determinism and Luddism. Rather, there is a need to find alternative ways of thinking about the relationship between humans and technology. As potential alternatives, he explored the physicist Gregory Stock's *Metahuman* (1993), who saw a need to build a metahuman to overcome the systems crisis, and the biologist James Lovelock's *Gaia* (1991), which offered an almost therapeutic vision of the globe in need of sensing and healing.

Extrapolating from the group's experience, I would like to believe that interpretive design research can study nonhumans from its perspective. The indirect route is obvious, even though it may be hard to learn. The theoretical route also exists, but the question is how to turn it into a working tool. Still, both routes require human intermediaries. By working with these intermediaries, interpretive design researchers can take steps toward constructing an antidote to Western humanism, which has subsumed the planet to human needs, without sacrificing its foundation of making sense of human beings.

Maybe this could be a base for an interpretive version of the "designing-with" that Wakkary is advocating—a design practice that binds designers together with natural and artificial environments by designing with tools developed by science to study nature and by engineering to make the human race less destructive to nature that supports it? Perhaps only time can tell whether this is enough. If the answer is no, we need a broad movement to create novel design methods that go beyond human-centered methods.

Interpretive Design Research and Imagination

These reflections take me to my final point. By recounting the history of one interpretive research program, I have tried to show that an interpretive approach can lead to significant progress over time and build toward an encompassing framework. I have also tried to show that it can also lead to collaborations that would have been hard to imagine only thirty years ago, and it is in fact still needed, even though design research has been moving away from a user-centered perspective over the last decade. Interpretive design researchers have faced a few obstacles working with colleagues whose goals they do not share, but they have overcome those barriers. For instance, craft designers have been interested in exploring the inner depths of their selves, sustainable designers in the well-being of the environment, and scientists in materials.

As we have seen, an interpretive approach can perhaps be extended to designing with nonhumans in mind by collaborating with specialists who know how to study nonhuman entities and processes. The key point to realize is that nonhumans are part of networks of meaning that shape their future. Many of these networks of meaning that are shaping the world in business, government, culture, and the sciences are destructive and serve no purpose nobler than making money, but some contain seeds for a future that is more sensitive to human and nonhuman needs. Interpretive design research can provide a way to collaborate with people who are forming these imaginaries and help to create alternatives from within—alternatives that are not driven by greed and power.

I am advocating extending the role of design as a mediator between art and technology, as in the 1920s, and as a mediator among art, technology, and science (including the social and human sciences) after the 1950s. Designers can grant agency to microbes, cotton fields, birds, and geological formations by creating processes and methods for studying them with microbiologists, agrologists, ornithologists, and geologists, just as they granted agency earlier to technology and society.

Yet there is inherent value in keeping humans in the picture. One thing that we should be mindful of is to avoid prioritizing either humans or nonhumans and to do everything we can to avoid building systems that start to dominate us. One of the maestros of Italian design was Tomás Maldonado. In his book *Design, Nature and Revolution* (1972), he outlined a dystopian vision of instrumental rationalism in Western culture. For him, we are living in prisons of our own creation and these prisons have the unprecedented capacity for evil:

> What is really happening today is that men are being transformed into things so that it will be easier to administer them. Instead of working with men, one can work with schemes, numbers, and graphs that represent men. In that context, models became more important than the objects of the persons of which they were a mere replica. For many years now, the fetishism of models, especially in the fields of economics, politics, and military strategy, has typified the attitude of the late

> Enlightenment of the modern technocrats. According to these people, perfection
> of the instructional and decision-making process is possible only if one succeeds in
> getting rid of all subjective interference with the construction and manipulation of
> the models used for obtaining that perfection . . . We have arrived at today's "post-
> Auschwitz" in part through the disappointing experience of attempting to make
> calculation an instrument of liberating violence. (Maldonado 1972, 20–22)

Maldonado finds the root cause of the crisis to be Western humanism,
which turns the planet into a servant of human needs. Our technological,
economic, and political systems have become means to serve humanity by
exploiting nature. They may be good for us in the short run, but they will
create an Auschwitz for the planet. We can find initial elements for such
design from Maldonado, who was talking about Buckminster Fuller's vision
of a rational, planned world devoid of politics and conflict. In contrast to
Fuller, who believed in the power of technology, Maldonado advocated a
philosophy that would revolutionize design by bringing together technical
imagination with the sociological imagination of C. Wright Mills, a polemi-
cal sociologist and intellectual hero of the 1950s. Creating a better world is an
act that requires both technical and political courage (Maldonado 1972, 29).

Not even Maldonado could see how badly Western humanism had dam-
aged the planet by 2020, however. It has led to environmental woes that were
unimaginable in the 1960s, and the scale of these woes is planetary. It is also
creating havoc in the human world. Flat-sharing services have turned some of
the world's most cherished cities into touristic theme parks, the web has turned
our world into yet another shopping mall, and face recognition systems are
turning parts of the world into massive prisons. Still, although humankind
may be the main source of the current planetary problems, it is also the solution. Were Maldonado still writing today, he might agree with Wakkary and
urge us to develop "nonhuman imagination" as part of his plea for a new
kind of design, but he would certainly encourage us to keep Wright's sociological imagination as a foundation. If this is so, interpretive ideals might be a
particularly good check against the forces that worried him. If we take human
beings seriously—to understand their theories and calculations, dreams and
hopes, fears and worries, changes of mind and obsessions, organizations and
institutions, fads and fashions, beliefs, and moral dilemmas—it becomes very
hard to push imposing top-down visions on them by saying that expertise justifies authority. Maybe interpretive design research could be a way to achieve
a new science of meaning that makes the blue marble of Earth a nurturing
environment for everyone.

References

Aakko, Maarit. 2016. *Fashion in-Between: Artisanal Design and Production of Fashion.* Helsinki: Aalto.

Aalto University. 2020. "Aalto University Bioinnovation Center Launching." https://www.aalto.fi/en/news/aalto-university-bioinnovation-centre-launching. Accessed January 8, 2022.

Acharya, Karthikeya. 2016. *Opening the Electrome: Redefining Home for Energy Studies through Design Practice.* Helsinki: Aalto.

Acharya, Karthikeya, Samir Bhowmik, and Jussi Mikkonen. 2013. "Light Is History." In *Proceedings of NORDES,* 511–512. Copenhagen: Royal Danish Academy of Fine Arts.

Ahde, Petra, and Jussi Mikkonen. 2008. "Hello: Bracelets Communicating Nearby Presence of Friends." In *Proceedings of the Tenth Anniversary Conference on Participatory Design,* 324–325. Indianapolis: University of Indiana.

Ahde, Petra, Jussi Mikkonen, and Sanna Latva-Ranta. 2009. "Spatial Jewelry: An Experiment with Communicative Party Jewelry." Poster presented at the IASDR Conference, Seoul, October 18–22.

Ahde-Deal, Petra. 2006. *Bling Bling—miten tytöt tykkäävät koruista* [Bling bling—how girls like jewelry]. Helsinki: Kuluttajatutkimuskeskus.

Ahde-Deal, Petra. 2013. *Women and Jewelry: A Social Approach to Wearing and Possessing Jewelry.* Helsinki: Aalto.

Ahonen, Annu, ed. 1996. *Code: 12 Styles 60 Small Domestic Appliances.* Helsinki: UIAH.

Alessi and UIAH. 1995. *The Workshop.* Milano: Nero su Bianco.

Alessi and UIAH. 2003. *Keittiössä: Taikkilaiset kokkaa Alessille—UIAH Students Cooking for Alessi.* Helsinki: UIAH.

Alexander, Christopher. 1964. *Notes on the Synthesis of Form.* Cambridge, MA: Harvard University Press.

Annicchiarico, Silvana, and Beppe Finessi. 2014. *Il design italiano oltre le crisi* [Italian design beyond the crisis]. Milan: Triennale Design Museum and Corraini Edizioni.

Appadurai, Arjun, ed. 1999. *The Social Life of Things: Commodities in Cultural Perspective.* London: Cambridge University Press. First published 1986 by Cambridge University Press (Cambridge, UK).

Arminen, Ilkka. 2004. *Institutional Interaction: Studies of Talk at Work.* Aldershot, UK: Ashgate.

Aspara, Jaakko. 2009. *Where Product Design Meets Investor Behavior: How Do Individual Investors' Evaluations of Companies' Product Design Influence Their Investment Decisions?* Helsinki: UIAH.

Bailey, F. G. 1983. *The Tactical Uses of Passion: An Essay on Power, Reason, and Reality.* Ithaca, NY, and London: Cornell University Press.

Ball, Roger. 2011. *SizeChina: A 3D Anthropometric Survey of the Chinese Head.* Delft, Netherlands: Delft University of Technology.

Bason, Christian. 2017. *Leading Public Design: How Managers Engage with Design to Transform Public Governance.* Copenhagen: Copenhagen Business School.

Battarbee, Katja. 2003a. "Co-Experience—the Social User Experience." In *Proceedings of Computer-Human Interaction CHI'03*, 730–731. New York: ACM.

Battarbee, Katja. 2003b. "Stories as Shortcuts to Meaning." In *Empathic Design*, edited by Ilpo Koskinen, Katja Battarbee, and Tuuli Mattelmäki, 107–118. Helsinki: IT Press.

Battarbee, Katja. 2004. *Co-Experience: Understanding User Experiences in Social Interaction.* Helsinki: UIAH.

Battarbee, Katja, and Ilpo Koskinen. 2004. "Co-Experience: User Experience as Interaction." *CoDesign* 1(1): 5–18.

Baudrillard, Jean. 1976. *L'échange symbolique et la mort* [Symbolic exchange and death]. Paris: Éditions Gallimard.

Bauman, Zygmunt. 1992. *Intimations of Postmodernity.* London: Routledge.

Beaver, Jacob, Tobie Kerridge, and Sarah Pennington. 2009. *Material Beliefs.* London: Goldsmiths, Interaction Research Studio.

Bell, Genevieve, Mark Blythe, and Phoebe Sengers. 2005. "Making by Making Strange: Defamiliarization and the Design of Domestic Technologies." *ACM Transactions on Computer-Human Interaction* 12: 149–173.

Bertola, Paola, ed. 2009. *Sistema Design Milano* [Milan Design System]. Milan: Abitare Segesta.

Beyer, Hugh, and Karen Holtzblatt. 1998. *Contextual Design: Defining Custom-Centered Systems*. San Francisco: Morgan Kaufmann.

Biggs, Michael. 2002. "The Role of the Artefact in Art and Design Research." *International Journal of Design Sciences and Technology* 10: 19–24.

Bijker, Wiebe, Thomas Hughes, and Trevor Pinch, eds. 1987. *The Social Construction of Technological Systems: New Directions in the Sociology and History of Technology*. Cambridge, MA: MIT Press.

Binder, Thomas. 2007. "Why Design: Labs." In *Proceedings of 2nd Nordic Design Research Conference (NORDES)*, May 27–30, Stockholm, Sweden: University of Arts, Craft, and Design. nordes.org.

Binder, Thomas, Giorgio De Michelis, Pelle Ehn, Per Linde, Giulio Jacucci, and Ina Wagner. 2011. *Design Things*. Cambridge, MA: MIT Press.

Black, Alison. 1998. "Empathic Design: User-Focused Strategies for Innovation." In *Proceedings of New Product Development, IBC Conferences*, 1–8. London: IBC.

Bleecker, Julian. 2009. *Design Fiction*. drbfw5wfjlxon.cloudfront.net/writing/Design Fiction_WebEdition.pdf.

Blumer, Herbert. 1954. "What Is Wrong with Social Theory?" *American Sociological Review* 18: 3–10.

Blumer, Herbert. 1969. *Symbolic Interactionism: Perspective and Method*. Berkeley: University of California.

Blythe, Mark. 2014. "Research through Design Fiction: Narrative in Real and Imaginary Abstracts." In *Proceedings of Conference on Human Factors in Computing Systems CHI'14*, 703–712. New York: ACM.

Blythe, Mark, Kees Overbeeke, Andrew Monk, and Peter Wright, eds. 2002. *Funology: From Usability to Enjoyment*. Dordrecht, Netherlands: Kluwer.

Borges, Adélia. 2011. *Design+Craft: The Brazilian Path*. São Paulo: Terceiro Nome.

Borgmann, Albert. 1995. "The Depth of Design." In *Discovering Design*, edited by Richard Buchanan and Victor Margolin, 13–22. Chicago: University of Chicago Press.

Boulding, Kenneth. 1989. *Economics as an Ecology of Commodities*. Lecture Notes at George Mason University, November, Fairfax, Virginia.

Branzi, Andrea. 1988. *Learning from Milan: Design and the Second Modernity*. Cambridge, MA: MIT Press.

Branzi, Andrea. 2008. *Che cosa è il design italiano? Le sette ossession* [What is Italian design? The seven obsessions]. Milan: La Triennale di Milano.

Brown, Tim. 2008. *Change by Design*. New York: HarperCollins.

Bruner, Jerome. 1990. *Acts of Meaning*. Cambridge, MA: Harvard University Press.

Buchanan, Richard. 2001. "Design Research and the New Learning." *Design Issues* 17: 3–23.

Buchenau, Marion, and Jane Fulton Suri. 2000. "Experience Prototyping." In *Proceedings of Designing Interactive Systems DIS'00*, 424–433. New York: ACM.

Butter, Reinhard. 1989. "Putting Theory into Practice: An Application of Product Semantics to Transportation Design." *Design Issues* 5: 51–67.

Buur, Jacob, and Henry Larsen. 2010. "The Quality of Conversations in Participatory Innovation." *CoDesign* 6: 121–138.

Buxton, Bill. 2007. *Sketching User Experiences: Getting the Design Right and the Right Design*. San Francisco: Morgan Kaufmann.

Cagan, Jonathan, and Craig M. Vogel. 2002. *Creating Breakthrough Products: Innovation from Product Planning to Program Approval*. Upper Saddle River, NJ: Prentice-Hall.

Caillois, Roger. 1961. *Man, Play and Games*. Translated by Meyer Barash. Urbana and Chicago: University of Illinois Press. First published as *Les Jeux et les Hommes: Le masque et le vertige* in French 1958 by Librairie Galliinard (Paris).

Cain, John. 1998. "Experience-Based Design: Towards a Science of Artful Business Innovation." *Design Management Journal* (Fall): 10–14.

Callon, Michel. 1987. "Society in the Making: The Study of Technology as a Tool for Sociological Analysis." In *The Social Construction of Technological Systems: New Directions in the Sociology and History of Technology*, edited by Wiebe Bijker, Thomas Hughes, and Trevor Pinch, 83–103. Cambridge, MA: MIT Press.

Campbell, Donald T. 1975. "'Degrees of Freedom' and the Case Study." *Comparative Political Studies* 8: 178–193.

Carroll, John. 2000. *Making Use: Scenario-Based Design of Human-Computer Interactions*. Cambridge, MA: MIT Press.

Chapman, Jonathan. 2005. *Emotionally Durable Design: Objects, Experiences and Empathy*. London: Earthscan.

Chapman, Jonathan. 2021. *Meaningful Stuff: Design That Lasts*. Cambridge, MA: MIT Press.

Chen, Dung-Sheng, Lu-lin Cheng, Caroline Hummels, and Ilpo Koskinen. 2016. "Social Design: An Introduction." *International Journal of Design* 10(1): 1–5.

Chun, Namkyu. 2018. *Re(dis)covering Fashion Designers: Interweaving Dressmaking and Placemaking*. Helsinki: Aalto.

Cohen, Michael D., James G. March, and Johan P. Olsen. 1972. "A Garbage Can Model of Organizational Choice." *Administrative Science Quarterly* 17: 1–25.

Colomina, Beatriz, and Mark Wigley. 2016. *Are We Human? Notes on an Archaeology of Design*. Zürich: Lars Müller Publishers.

Crabtree, Andy. 2003. *Designing Collaborative Systems: A Practical Guide to Ethnography*. London: Springer.

Crabtree, Andy. 2004. "Design in the Absence of Practice: Breaching Experiments." In *Proceedings of Designing Interactive Systems DIS'04*, 59–84. New York: ACM.

Cross, Nigel. 2001. "Designerly Ways of Knowing. Design Discipline Versus Design Science." *Design Issues* 17(3): 49–55.

Csikszentmihalyi, Mihaly, and Eugene Rochberg-Halton. 1981. *The Meaning of Things: Domestic Symbols and the Self*. Cambridge: Cambridge University Press.

Dalsgaard, Peter. 2009. *Designing Engaging Interactive Environments: A Pragmatist Perspective*. Aarhus, Denmark: Aarhus University.

Dandavate, Uduy, Elizabeth B.-N. Sanders, and Susan Stuart. 1996. "Emotions Matter: User Empathy in the Product Development Process." In *Proceedings of the Human Factors and Ergonomics Society Annual Meeting* 40(7), 415–418. doi:10.1177/154193129604000709.

Dewey, John. 1980. *Art as Experience*. New York: Perigee Books.

Dixon, Brian. 2020. *Dewey and Design*. Dordrecht, Netherlands: Springer.

Djajadiningrat, Tom. 1998. *Cubby: What You See Is Where You Act*. Delft, Netherlands: Delft University of Technology.

Djajadiningrat, Tom, Kees Overbeeke, and Stephan Wensveen. 2002. "But How, Donald, Tell Us How? On the Creation of Meaning in Interaction Design through Feedforward and Inherent Feedback." In *Proceedings of Designing Interactive Systems DIS'02*, 285–291. New York: ACM.

Dourish, Paul. 2002. *Where the Action Is*. Cambridge, MA: MIT Press.

Dumas, Joe. 2007. "The Great Leap Forward: The Birth of the Usability Profession (1988–1993)." *Journal of Usability Studies* 2: 54–60.

Dunne, Anthony. 2005. *Hertzian Tales*. Cambridge, MA: MIT Press. First published 1999 by RCA (London).

Dunne, Anthony, and William Gaver. 2001. *Presence Project*. RCA CRD Projects Series. London: RCA.

Dunne, Anthony, and Fiona Raby. 2001. *Design Noir: The Secret Life of Electronic Objects*. Basel: August/Birkhäuser.

Dunne, Anthony, and Fiona Raby. 2013. *Speculative Everything: Design, Fiction, and Social Dreaming.* Cambridge, MA: MIT Press.

Durkheim, Émile. 1980. *Uskontoelämän alkeismuodot: Australialainen toteemijärjestelmä* [The Elementary Forms of Religious Life]. Translated by Seppo Randell. Helsinki: Tammi. First published as *Les formes élémentaires de la vie religieuse* in French 1912 by F. Alcan (Paris).

Durrani, Marium. 2019. *Through the Threaded Needle: A Multi-Sited Ethnography on the Sociomateriality of Garment Mending Practices.* Helsinki: Aalto.

DWoC. 2016. *Design Driven Value Chains in the World of Cellulose.* https://chemarts.aalto.fi/wp-content/uploads/2015/05/1_DWoC_final_presentation_20052015.pdf.

Dyer, Jeff, Hal B. Gregersen, and Clayton Christensen. 2011. *The Innovator's DNA: Mastering the Five Skills of Disruptive Innovators.* Boston: Harvard Business Review Press.

Eerola, Markus. 1999. *Värillisiä pintoja lasissa. Liekkiruiskutusmenetelmän kehittäminen lasiesineiden valmistamiseksi* [Color surfaces in glass: Developing a flame-spraying method for glassware]. Helsinki: UIAH.

Ehn, Pelle. 1988. *Work-Oriented Design of Computer Artifacts.* Stockholm: Arbetslivscentrum.

Ehn, Pelle. 1998. "Manifesto for a Digital Bauhaus." *Digital Creativity* 9: 207–216.

Ehn, Pelle, and Morten Kyng. 1991. "Cardboard Computers: Mocking-It-Up or Hands-On the Future." In *Design at Work: Cooperative Design of Computer Systems*, edited by Joan Greenbaum and Morten Kyng, 169–196. Hillsdale, NJ: Lawrence Erlbaum.

Ericson, Magnus, Martin Frostner, Zac Kyes, Sarah Teleman, and Jonas Williamson, eds. 2009. *Iaspis Forum on Design and Critical Practice.* Stockholm: Iaspis and Sternberg Press.

Ernevi, Anders, Samuel Palm, and Johan Redström. 2005. "Erratic Appliances and Energy Awareness." In *Proceedings of Nordic Design Research Conference NORDES*, May 29–31, Copenhagen. nordes.org/.

Forlano, Laura. 2017. "Posthumanism and Design." *She Ji* 3: 16–29.

Forlizzi, Jodi, and Katja Battarbee. 2005. "Understanding Experience in Interactive Systems." In *Proceedings of Designing Interactive Systems DIS'04*, 261–268. New York: ACM Press.

Forlizzi, Jodi, and Shannon Ford. 2000. "The Building Blocks of Experience: An Early Framework for Interaction Designers." In *Proceedings of Designing Interactive Systems DIS'00*, 419–423. New York: ACM.

Frampton, Kenneth. 1983. "Towards a Critical Regionalism: Six Points for an Architecture of Resistance." In *The Anti-Aesthetic: Essays on Postmodern Culture*, edited by Hal Foster, 16–30. Seattle: Bay Press.

Frascara, Jorge, ed. 2002. *Design and the Social Sciences: Making Connections*. London: Taylor & Francis.

Frayling, Christopher. 1993. *Research in Art and Design*. London: Royal College of Art.

Freire, Paolo. 2005. *Pedagogy of the Oppressed*. 30th anniversary ed. Translated by M. B. Ramos. New York: Continuum. First published as *Pedagogia do Oprimido* in Portuguese 1972 by Afrontamento (Porto, Portugal).

Frens, Joep. 2006. *Designing for Rich Interaction: Integrating Form, Interaction, and Function*. Eindhoven, Netherlands: Technishe Universiteit Eindhoven.

Fuad-Luke, Alistair. 2009. *Design Activism: Beautiful Strangeness for a Sustainable World*. London: Earthscan.

Fukasawa, Naoto, and Jasper Morrison. 2007. *Super Normal: Sensations of the Ordinary*. Zürich: Lars Müller.

Fulton, Jane. 1993. "Physiology and Design: Ideas about Physiological Human Factors and the Consequences for Design Practice." *American Center for Design Journal* 7: 7–15.

Fulton Suri, Jane. 2003. "Empathic Design: Informed and Inspired by Other People's Experience." In *Empathic Design*, edited by Ilpo Koskinen, Katja Battarbee, and Tuuli Mattelmäki, 51–58. Helsinki: IT Press.

Garduño García, Claudia. 2017. *Design as Freedom*. Helsinki: Aalto.

Garfinkel, Harold. 1967. *Studies in Ethnomethodology*. Englewood Cliffs, NJ: Prentice-Hall.

Gaver, Bill, Tony Dunne, and Elena Pacenti. 1999. "Design: Cultural Probes." *Interactions* 6(1): 21–29.

Gaver, William. 2002. "Presentation about Cultural Probes." PowerPoint presentation, UIAH, Helsinki, November. smart.uiah.fi/luotain/pdf/probes-seminar/GaverPROBES.pdf/.

Gaver, William. 2012. "What Should We Expect from Research Through Design?" In *Proceedings of Conference on Human Factors in Computing Systems CHI'12*, 937–946. New York: ACM.

Gedenryd, Henrik. 1998. *How Designers Work: Cognitive Studies*. Lund, Sweden: Lund University. lucs.lu.se/People/Henrik.Gedenryd/HowDesignersWork/.

Geertz, Clifford. 1973. "Thick Description: Towards an Interpretive Theory of Culture." In *The Interpretation of Cultures*, 3–30. New York: Basic Books.

Giaccardi, Elisa, and Johan Redström. 2020. "Technology and More-Than-Human Design." *Design Issues* 36: 33–44.

Gladwell, Malcolm. 1997. "The Coolhunt." http://gladwell.com/the-coolhunt/.

Glaser, Barney G., and Anselm L. Strauss. 1967. *The Discovery of Grounded Theory: Strategies for Qualitative Research*. New York: Aldine de Gruyter.

Goffman, Erving. 1961. *Encounters: Two Studies in the Sociology of Interaction—Fun in Games and Role Distance*. Indianapolis: Bobbs-Merrill.

Goffman, Erving. 2009. *Stigma: Notes on the Management of Spoiled Identity*. London: Penguin Books. First published 1963 by Penguin Books (London).

Gouldner, Alvin. 1970. *The Coming Crisis of Western Sociology*. London: Heinemann.

Graves Petersen, Marianne, Ole Iversen, Peter Gall Krogh, and Martin Ludvigsen. 2004. "Aesthetic Interaction: A Pragmatist's Aesthetics of Interactive Systems." In *Proceedings of Designing Interactive Systems DIS'04*, 269–276. New York: ACM.

Grcic, Konstantin. 2010. *Design Real*. London: Koenig Books and Serpentine Gallery. Exhibition catalog.

Greenson, Ralph. 1967. *The Technique and Practice of Psycho-Analysis*. London: Hogarth Press and Institute of Psycho-Analysis.

Hakio, Kirsi, and Tuuli Mattelmäki. 2011. "Design Adventures in the Public Sector." In *Proceedings of DPPI Conference*, 1–8. New York: ACM.

Halle, David. 1993. *Inside Culture: Art and Class in the American Home*. Chicago: University of Chicago Press.

Hamel, Gary. 2002. *Leading the Revolution: How to Thrive in Turbulent Times by Making Innovation a Way of Life*. New York: Penguin.

Hamel, Gary, and C. K. Prahalad. 1994. *Competing for the Future*. Boston: Harvard Business School Press.

Haraway, Donna. 1988. "Situated Knowledges: The Science Question in Feminism and the Privilege of Partial Perspective." *Feminist Studies* 14(3): 575–599.

Härkäsalmi, Tiina. 2008. *Runkokuituja lyhytkuitumenetelmin: Kohti pellavan ja hampun ympäristömyötäistä tuotteistamista* [Bast fibers by short-fibre methods: Towards an environmentally conscious productization of flax and hemp]. Helsinki: Taik.

Harper, Richard, ed. 2003. *Inside the Smart Home*. London: Springer.

Hassenzahl, Marc. 2004. "The Interplay of Beauty, Goodness, and Usability in Interactive Products." *Human-Computer Interaction* 19: 319–349.

Heidegger, Martin. 2010. *Being and Time*. Translated by J. Stambaugh. Albany: State University of New York Press. First published as *Sein und Zeit* in German 1927 by Max Niemeyer Verlag (Halle, Germany).

Heljakka, Katriina. 2013. *Principles of Adult Play(fulness) in Contemporary Toy Cultures: From Wow to Flow to Glow*. Helsinki: Aalto.

Heskett, John 1989. *Philips: A Study of the Corporate Management of Design*. London: Trefoil.

Hevner, Alan, Sudna Ram, Salvatore March, and Jinsoo Park. 2004. "Design Science in Information Systems Research." *MIS Quarterly* 28: 75–105.

Hirscher, Anja-Lisa. 2020. *When Skillful Participation Becomes Design: Making Clothes Together*. Helsinki: Aalto.

Hirschheim, Rudy, and Heinz K. Klein. 1989. "Four Paradigms of Information Systems Development." *Communications of the ACM* 32: 1199–1216.

Hochschild, Arlie R. 2003. *The Managed Heart: Commercialization of Human Feelings*. Berkeley: University of California Press.

Hofmeester, Kay, and Esther de Charon de Saint Germain, eds. 1999. *Presence: New Media for Older People*. Amsterdam: Netherlands Design Institute.

Hofmeester, Kay, and Yvon Gijsbers, eds. 1999. "The Digital Hug: Families Keeping in Touch." Special issue, *ACM Interactions* 6, no. 6 (November/December).

Huhtamaa, Inkeri. 2010. *Namibian Bodily Appearance and Handmade Objects: The Meanings of Appearance Culture and Handmade Objects from the Perspective of the Craft Persons*. Helsinki: Aalto.

Huizinga, Johan. 1950. *Homo Ludens: A Study of the Play Element in Culture*. London: Routledge & Kegan Paul.

Hummels, Caroline. 2000. *Gestural Design Tools: Prototypes, Experiments and Scenarios*. Delft, Netherlands: Delft University of Technology.

Hutchins, Edwin. 1996. *Cognition in the Wild*. Cambridge, MA: MIT Press.

Iacucci, Giulio, Kari Kuutti, and Mervi Ranta. 2000. "On the Move with a Magic Thing. Role-Playing in Concept Design of Mobile Services and Devices." In *Proceedings of Designing Interactive Systems DIS'00*, 193–202. New York: ACM.

Ihde, Don. 2006. "The Designer Fallacy and Technological Imagination." In *Defining Technological Literacy*, edited by John Dakers, 121–131. New York: Palgrave Macmillan.

IP08 (Ilpo Koskinen, Jussi Mikkonen, Petra Ahde, Kaj Eckoldt, Thorsteinn Helgason, Riikka Hänninen, Jing Jiang, Timo Niskanen, and Benjamin Schultz). 2009. "Hacking a Car: Re-embodying the Design Classroom." In *Proceedings of Nordic Design Research Conference NORDES'09*, August 30–September 1, Oslo, Norway: The Oslo School of Architecture and Design. nordes.org.

Issey Miyake. 2016. *Miyake Issey Exhibition: The Work of Miyake Issey*. Tokyo: National Art Center. Exhibition catalog. https://www.nact.jp/english/exhibitions/2016/miyake_is

sey/#:~:text=An%20exhibition%20devoted%20to%20designer,the%20National%20
Art%20Center%2C%20Tokyo.

Jacobson, Susanne. 2014. *Personalised Assistive Products: Managing Stigma and Expressing the Self*. Helsinki: Aalto.

Jacucci, Giulio. 2004. *Interaction as Performance: Cases of Configuring Physical Interfaces in Mixed Media*. Oulu, Finland: Oulu University Press.

Järvinen, Juha, and Ilpo Koskinen. 2001. *Industrial Design as a Culturally Reflexive
Activity in Manufacturing*. Helsinki: UIAH and Sitra.

Jensen, Rolf. 1999. *The Dream Society*. New York: McGraw-Hill.

Johansson, Sofia, Pete Kaario, Anu Kankainen, Vesa Kantola, Mikael Runonen, and
Kirsikka Vaajakallio. 2010. *eXtreme Design Final Report*. https://designresearch.aalto.fi
/groups/encore/wp-content/uploads/2012/06/exdesign_final_report.pdf.

Jordan, Patrick. 1998. "Human Factors for Pleasure in Product Use." *Applied Ergonomics* 29: 25–33.

Jordan, Patrick. 2000. *Designing Pleasurable Products*. London: Taylor and Francis.

Jouet, Jacques. 2000. *Poèmes de métro*. Paris: P.O.L.

Júdice, Andrea. 2014. *Design for Hope: Designing Health Information in Vila Rosário*.
Helsinki: Aalto.

Júdice, Andrea, Marcelo Júdice, and Ilpo Koskinen. 2015. "Enriching Ethnography
in Marginalized Communities with Surrealist Techniques." *Ethnography Praxis in
Industry Conference Proceedings* (1): 119–131. https://anthrosource.onlinelibrary.wiley
.com/doi/pdf/10.1111/1559-8918.2015.01044.

Júdice, Marcelo. 2014. *You Are Important! Empowering Health Agents in Vila Rosário
through Design*. Helsinki: Aalto.

Kääriäinen, Pirjo, and Liisa Tervinen, eds. 2017. *Lost in the Wood(s)*. Helsinki:
Aalto.

Kääriäinen, Pirjo, Liisa Tervinen, Tapani Vuorinen, and Eeva Suorlahti, eds. 2020.
The CHEMARTS Cookbook. Helsinki: Aalto.

Kankainen, Anu, Kirsikka Vaajakallio, Vesa Kantola, and Tuuli Mattelmäki. 2012.
"Storytelling Group. A Co-Design Method for Service Design." *Behavior and Information Technology* 31: 221–230.

Kanter, Rosabeth Moss. 1972. *Commitment and Community: Communes and Utopias in
Sociological Perspective*. Cambridge, MA: Harvard University Press.

Karjalainen, Toni-Matti. 2004. *Semiotic Transformations in Design: Communicating
Strategic Product Identity through Product Design References*. Helsinki: UIAH.

Kasanen, Eero, Kari Lukka, and Arto Siitonen. 1993. "The Constructive Approach in Management Accounting Research." *Journal of Management Accounting Research* 5: 241–246.

Kataja, Kirsi, and Pirjo Kääriäinen, eds. 2018. *Designing Cellulose for the Future: Design-Driven Value Chains in the World of Cellulose (DWoC) 2013–2018.* Helsinki: Copy-Set.

Katz, Jack. 1999. *How Emotions Work.* Chicago: University of Chicago Press.

Keinonen, Turkka. 1998. *One-Dimensional Usability: Influence of Usability on Consumers' Product Preference.* Helsinki: University of Arts and Design.

Keinonen, Turkka, and Milvi Soosalu. 1997. *Puzzle Interview: Visual Support for User Involvement.* Helsinki: University of Art and Design.

Kelley, Tom. 2001. *The Art of Innovation: Lessons in Creativity from IDEO, America's Leading Design Firm.* New York: Random House.

Kemper, Theodore D. 1981. "Social Constructionist and Positivist Approaches to the Sociology of Emotions." *American Journal of Sociology* 87: 336–362.

Kester, Grant. 2004. *Conversation Pieces: Communication and Community in Modern Art.* Berkeley: University of California Press.

Kester, Grant. 2011. *The One and the Many: Contemporary Collaborative Art in a Global Context.* Durham, NC: Duke University Press.

Kicherer, Sibylle. 1990. *Olivetti: A Study of the Corporate Management of Design.* London: Trefoil.

Kohtala, Cindy. 2016. *Making Sustainability: How Fab Labs Address Environmental Issues.* Helsinki: Aalto.

Kopytoff, Igor. 1986. "The Cultural Biography of Things: Commodization as Process." In *The Social Life of Things, Commodities in Cultural Perspective*, edited by Arjun Appadurai, 64–92. London: Cambridge University Press.

Korhonen, Panu. 2000. "Käytettävyystesteistä liiketoiminnan ytimeen" [From usability tests to the core of the business]. In *Miten käytettävyys muotoillaan?*, edited by Turkka Keinonen, 181–192. Helsinki: Taik.

Koskinen, Ilpo. 2003. "Empathic Design in Methodic Terms." In *Empathic Design*, edited by Ilpo Koskinen, Katja Battarbee, and Tuuli Mattelmäki, 59–65. Helsinki: IT Press.

Koskinen, Ilpo. 2007. *Mobile Multimedia in Action.* New Brunswick, NJ: Transaction.

Koskinen, Ilpo. 2016. "Agonist, Convivial and Conceptual Aesthetic in Social Design." *Design Issues* 32: 18–29.

Koskinen, Ilpo, and Katja Battarbee. 2003. "Preface." In *Empathic Design*, edited by Ilpo Koskinen, Katja Battarbee, and Tuuli Mattelmäki, 1–11. Helsinki: IT Press.

Koskinen, Ilpo, Katja Battarbee, and Tuuli Mattelmäki, eds. 2003. *Empathic Design*. Helsinki: IT Press.

Koskinen, Ilpo, Thomas Binder, and Johan Redström. 2008. "Lab, Field, Gallery, and Beyond." *Artifact: Journal of Virtual Design* 2: 46–57.

Koskinen, Ilpo, Thomas Binder, Johan Redström, Stephan Wensveen, and John Zimmerman. 2011. *Design Research through Practice: From Lab, Field, and Showroom*. San Francisco: Morgan Kaufmann.

Koskinen, Ilpo, Jodi Forlizzi, and Katja Battarbee. Forthcoming. "Expanding Pragmatism with Symbolic Interactionism: Recounting the Story of Two Frameworks." *Design Issues* 39(4).

Koskinen, Ilpo, and Heidi Grönman. 2006. *Design and Domestication: Presented at Designing Consuming*. Glasgow: Glasgow University.

Koskinen, Ilpo, Esko Kurvinen, and Turo-Kimmo Lehtonen. 2002. *Mobile Image*. Helsinki: IT Press.

Koskinen, Ilpo, Kristo Kuusela, Katja Battarbee, Anne Soronen, Frans Mäyrä, Jussi Mikkonen, and Mari Zakrzewski. 2006. "Morphome: A Field Study of Proactive Information Technology at Home." In *Proceedings of Designing Information Systems DIS'06*, 179–188. New York: ACM.

Kosonen, Krista. 2018. *Finding One's Way in Design: Reflections on Narrative Professional Identity*. Helsinki: Aalto.

Krippendorff, Klaus. 2006. *The Semantic Turn: A New Foundation for Design*. Boca Raton, FL: Taylor & Francis.

Krogh, Peter, and Ilpo Koskinen. 2020. *Drifting by Intention: Four Epistemic Traditions from within Constructive Design Research*. Dordrecht, Netherlands: Springer.

Krogh, Peter, and Ilpo Koskinen. 2022. "Four Epistemologies of Constructive Design Research." *Design Issues* 38: 33–46.

Kurvinen, Esko. 2007. *Prototyping Social Action*. Helsinki: UIAH.

Kurvinen, Esko, Katja Battarbee, and Ilpo Koskinen. 2008. "Prototyping Social Interaction." *Design Issues* 21: 46–57.

Kuusk, Kristi, Marjan Kooroshnia, and Jussi Mikkonen. 2015. "Crafting Butterfly Lace: Conductive Multi-Color Sensor-Actuator Structure." In *Adjunct Proceedings of the 2015 ACM International Joint Conference on Pervasive and Ubiquitous Computing and Proceedings of the 2015 ACM International Symposium on Wearable Computers*, 595–600. New York: ACM.

Kuutti, Kari. 1996. "Activity Theory as a Potential Framework for Human-Computer Interaction Research." In *Context and Consciousness: Activity Theory and*

Human-Computer Interaction, edited by Bonnie A. Nardi, 9–22. Cambridge, MA: MIT Press.

Kwak, Youn-Jung, Tiia Suomalainen, and Jussi Mikkonen. 2011. "Study of Honest Signal: Bringing Unconscious Channel of Communication into Awareness through Interactive Prototype." In *Proceedings of the 2nd International Conference on Human Centered Design*, 529–536. Berlin and Heidelberg: Springer-Verlag.

Kymäläinen, Tiina. 2015. *Science Fiction Prototypes as Design Outcome of Research: Reflecting Ecological Research Approach and Experience Design for the Internet of Things.* Helsinki: Aalto.

Laine, Meri. 2001. "Sokkotreffit: Kokemuksia mukana kulkevien esineiden muodosta ja haptisuudesta" [Blind date: The experience of shape and hapticity of portable objects]. Master's thesis, UIAH, Helsinki.

Lakatos, Imre. 1970. "Falsification and the Methodology of Scientific Research Programmes." In *Criticism and the Growth of Knowledge*, edited by Imre Lakatos and Alan Musgrave, 8–101. Cambridge, UK: Cambridge University Press.

Latour, Bruno. 1987. *Science in Action.* Cambridge, MA: Harvard University Press.

Latour, Bruno. 1990. "Technology Is Society Made Durable." *The Sociological Review*, 38(1_suppl): 103–131. https://doi.org/10.1111/j.1467-954x.1990.tb03350.x.

Laurel, Brenda. 1991. *Computers as Theater.* Upper Saddle River, NJ: Addison-Wesley.

Laurel, Brenda, ed. 2003. *Design Research: Methods and Perspectives.* Cambridge, MA: MIT Press.

Lee, Jung-Joo. 2012. *Against Method: The Portability of Method in Human-Centered Design.* Helsinki: Aalto.

Lee, Jung-Joo, Jack Whalen, and Ilpo Koskinen. 2021. "Multiple Intelligibility in Constructive Design Research." *International Journal of Design* 14(3): 55–66.

Lenay, Charles. 2010. "'It's So Touching': Emotional Value in Distal Contact." *International Journal of Design* 4: 15–25.

Leonard, Dorothy, and Jeffrey E. Rayport. 1997. "Spark Innovation through Empathic Design." *Harvard Business Review* (November–December): 102–113.

Leppänen, Helena. 2006. *Muotoilija ja toinen: Astiasuunnittelua vanhuuden kontekstissa* [The designer and the other: Designing dishes in the context of old age]. Helsinki: Taik.

Lettenmeier, Michael. 2018. *A Sustainable Level of Material Footprint: Benchmark for Designing One-Planet Lifestyles.* Helsinki: Aalto.

Lewin, Kurt. 1946. "Action Research and Minority Problems." *Journal of Social Issues* 2: 34–46.

Lindley, Joseph, and Paul Coulton. 2016. "Pushing the Limits of Design Fiction: The Case for Fictional Research Papers." In *Proceedings of Conference on Human Factors in Computing Systems CHI'16*, 4032–4043. New York: ACM.

Loewy, Raymond. 2002. *Never Leave Well Enough Done*. Baltimore and London: Johns Hopkins University Press. First published 1951 by Simon & Schuster (New York).

Lovell, Sophie. 2009. *Limited Edition: Prototypes, One-Offs and Design Art Furniture*. Basel, Switzerland: Birkhäuser.

Lovelock, James. 1991. *Gaia: A New Look at Life on Earth*. Oxford: Oxford University Press.

Ludvigsen, Martin. 2006. *Designing for Social Interaction*. Aarhus, Denmark: Aarhus School of Architecture.

Lykke-Olesen, Andreas. 2006. *Space as Interface: Bridging the Gap with Cameras*. Aarhus, Denmark: Aarhus School of Architecture.

Lynch, Michael. 1993. *Scientific Practice and Ordinary Action: Ethnomethodology and Social Studies of Science*. Cambridge, UK: Cambridge University Press.

Mäkelä, Anu, and Katja Battarbee. 1999. "It's Fun to Do Things Together: Two Cases of Explorative User Studies." *Personal Technologies* 3: 137–140.

Mäkelä, Anu, Verena Giller, Manfred Tscheligi, and Reinhard Sefelin. 2000. "Joking, Storytelling, Artsharing, Expressing Affection: A Field Trial of How Children and Their Social Network Communicate with Digital Images in Leisure Time." In *Proceedings of Conference on Human Factors in Computing Systems CHI'00*, 548–555. New York: ACM.

Mäkelä, Maarit. 2003. *Saveen piirtyviä muistoja* [Memories in clay]. Helsinki: Aalto.

Mäkelä, Maarit, and Sara Routarinne, eds. 2007. *The Art of Research: Research Practices in Art and Design*. Helsinki: UIAH.

Maldonado, Tomás. 1972. *Design, Nature, and Revolution: Toward a Critical Ecology*. New York: Harper & Row.

Manzini, Ezio. 2015. *Design, When Everybody Designs: An Introduction to Design for Social Innovation*. Cambridge, MA: MIT Press.

Marttila, Tatu. 2018. *Platforms of Co-Creation: Learning Interprofessional Design Practice in Creative Sustainability*. Helsinki: Aalto.

Maslow, Abraham. 1946. "A Theory of Human Motivation." *Psychological Review* 50: 370–396.

Maspéro, François. 1990. *Les passagers du Roissy-Express*. Paris: Seuil.

Mattelmäki, Tuuli. 2003. "Probes: Studying Experiences for Design Empathy." In *Empathic Design*, edited by Ilpo Koskinen, Katja Battarbee, and Tuuli Mattelmäki, 119–130. Helsinki: IT Press.

Mattelmäki, Tuuli. 2006. *Design Probes*. Helsinki: UIAH.

Mattelmäki, Tuuli, and Katja Battarbee. 2002. "Empathy Probes." In *Proceedings of the Participatory Design Conference*, 266–271. Palo Alto, CA: CPSR.

Mattelmäki, Tuuli, Eva Brandt, and Kirsikka Vaajakallio. 2011. "On Designing Open-Ended Interpretations for Collaborative Design Exploration." *CoDesign: International Journal of CoCreation in Design and the Arts* 2: 79–93.

Mattelmäki, Tuuli, and Kirsi Hakio. 2006. "Designing Alternative Arrangements for Ageing Workers." In *Proceedings of the Participatory Design Conference, Vol II*, 101–104. Palo Alto, CA: CPSR.

Mattelmäki, Tuuli, and Turkka Keinonen. 2001. "Design for Brawling: Exploring Emotional Issues for Concept Design." In *Proceedings of the International Conference of Affective Human Factors Design*, 148–155. London: Asean Academic Press.

Mattelmäki, Tuuli, Sara Routarinne, and Salu Ylirisku. 2011. "Triggering the Storytelling Mode." In *Proceedings of the Participatory Innovation Conference PINC'11*, 38–44. Sonderborg: University of Southern Denmark.

Mattelmäki, Tuuli, and Froukje Sleeswijk Visser. 2011. "Lost in Co-X: Interpretations of Co-Design and Co-Creation." In *Proceedings of IASDR 2011, the 4th World Conference on Design Research*, ed. N. Roozenburg, L. L. Chen, and P. J. Stappers, 1–12. Delft, Netherlands: Delft University of Technology/IASDR.

Mattelmäki, Tuuli, and Kirsikka Vaajakallio. n.d. *SPICE: Understanding Public Spaces through Narrative Concept Design*. Helsinki: Aalto.

Mattelmäki, Tuuli, Kirsikka Vaajakallio, and Ilpo Koskinen. 2014. "What Happened to Empathic Design?" *Design Issues* 30: 67–77.

Maxwell, John C. 2007. *Failing Forward: Turning Mistakes into Stepping Stones for Success*. Nashville: Thomas Nelson.

Mäyrä, Frans, Anne Soronen, Jukka Vanhala, Ilpo Koskinen, Kristo Kuusela, Jussi Mikkonen, and Mari Zakrzewski. 2006. "Probing a Proactive Home: Challenges in Researching and Designing Everyday Smart Environments." *Human Technology* 2: 158–186.

McCall, George J., and Jerry L. Simmons. 1978. *Identities and Interactions*. New York: Free Press.

McCarthy, John, and Peter Wright. 2004. *Technology as Experience*. Cambridge, MA: MIT Press.

McIntyre, Jean. 1995. "The Department of Design Research at the Royal College of Art: Its Origins and Legacy 1959–1988." In *Design of the Times: One Hundred Years of the Royal College of Art*, edited by Christopher Frayling, 59–62. Somerset, UK: Shepton Beauchamp.

Mead, George Herbert. 1934. *Mind, Self, and Society: From the Standpoint of a Social Behaviorist*. Chicago: University of Chicago Press.

Meroni, Anna, Daniela Selloni, and Martina Rossi. 2018. *Massive Co-Design: A Proposal for a Collaborative Design Framework*. Milan: FrancoAngeli.

Miettinen, Satu. 2007. *Designing the Creative Tourism Experience: A Service Design Process with Namibian Craftspeople*. Helsinki: UIAH.

Mikkonen, Jussi. 2016. *Prototyping Interactions*. Tampere, Finland: Tampere University of Technology.

Mikkonen, Jussi, Ramyah Gowrishankar, Miia Oksanen, Harri Raittinen, and Arto Kolinummi. 2014. "OJAS: Open Source Bi-Directional Inductive Power Link." In *Proceedings of the Conference on Human Factors in Computing Systems CHI'14*, 1049–1058. New York: ACM.

Mikkonen, Jussi, and Riitta Townsend. 2019. "Frequency-Based Design of Smart Textiles." In *Proceedings of Conference on Human Factors in Computing Systems CHI'19*, 1–12. New York: ACM. https://doi.org/10.1145/3290605.3300524.

Miller, Daniel, and Don Slater. 2000. *The Internet: An Ethnographic Approach*. Oxford: Berg.

Mogensen, Preben. 1992. "Towards a Provotyping Approach in Systems Development." *Scandinavian Journal of Information Systems* 4: 31–53.

Moggridge, Bill. 2006. *Designing Interactions*. Cambridge, MA: MIT Press.

Mumford, E., R. Hirschheim, G. Fitzgerald, and A.-T. Wood-Harper, eds. 1985. *Research Methods in Information Systems*. Amsterdam: North-Holland Publishing.

Nardi, Bonnie, ed. 1996. *Context and Consciousness: Activity Theory and Human-Computer Interaction*. Cambridge, MA: MIT Press.

Negroponte, Nicholas. 1995. *Being Digital*. New York: Vintage Books.

Niedderer, Kristina. 2004. "Designing the Performative Object: A Study in Designing Mindful Interaction through Artefacts." PhD diss., University of Plymouth, UK.

Nielsen, Jakob. 1993. *Usability Engineering*. San Francisco: Academic Press.

Nielsen, Jakob, and Rolf Molich. 1990. "Heuristic Evaluation of User Interfaces." In *Proceedings of Conference on Human Factors in Computing Systems CHI'90*, 249–256. New York: ACM.

Nieminen-Sundell, Riitta, and Mika Pantzar. 2003. "Towards an Ecology of Goods: Symbiosis and Competition Between Material Household Commodities." In *Empathic Design*, edited by Ilpo Koskinen, Katja Battarbee, and Tuuli Mattelmäki, 131–142. Helsinki: IT Press.

Niinimäki, Kirsi. 2011. *From Disposable to Sustainable: The Complex Interplay between Design and Consumption of Textiles and Clothing.* Helsinki: Aalto.

Norman, Don A. 1998. *The Invisible Computer: Why Good Products Can Fail, the Personal Computer Is so Complex, and Information Appliances Are the Solution.* Cambridge, MA: MIT Press.

Nugent, Lisa, Sean Donahue, Mia Berberat, Yee Chan, Justin Gier, Ilpo Koskinen, and Tuuli Mattelmäki. 2007. "How Do You Say Nature? Opening the Design Space with a Knowledge Environment." *Knowledge, Technology and Policy* 4: 269–279.

Ochs, Elinor, Sally Jacoby, and Patrick Gonzales. 1994. "Interpretive Journeys: How Physicists Talk and Travel through Graphic Space." *Configurations* 1: 151–171.

Osaki, Miya. 2008. "Retellings." Master's thesis, ArtCenter College of Design, Pasadena, CA. people.artcenter.edu/~osaki/retellings/index.html.

Overbeeke, Kees. 2007. "The Aesthetics of the Impossible." Inaugural lecture, Technische Universiteit Eindhoven, Netherlands. www.tue.nl/bib/.

Overbeeke, Kees, Stephan Wensveen, and Caroline Hummels. 2006. "Design Research: Generating Knowledge through Doing." In *Drawing New Territories: State of the Art and Perspectives. Third Symposium of Design Research*, 51–69. Geneva: Swiss Design Network.

Paavilainen, Heidi. 2014. *Dwelling with Design.* Helsinki: Aalto.

Pallasmaa, Juhani. 2009. *The Thinking Hand: Existential and Embodied Wisdom in Architecture.* Chichester, UK: Wiley.

Pantzar, Mika. 1991. *A Replicative Perspective on Evolutionary Dynamics: The Organizing Process of the US Economy Elaborated through Biological Metaphor.* Research Report 37/1991. Helsinki: Labour Institute for Economic Research.

Pantzar, Mika. 1993. "Do Commodities Reproduce themselves through Human Beings?" *World Futures—the Journal of General Evolution* 38: 201–224.

Pantzar, Mika. 1996. *Kuinka teknologia kesytetään* [How technology is domesticated]. Helsinki: Hanki ja Jää.

Pantzar, Mika. 2000. *Tulevaisuuden koti: Arjen tarpeita keksimässä* [Home of the future: Inventing everyday needs]. Helsinki: Otava.

Pantzar, Mika, and Elizabeth Shove, eds. 2005. *Manufacturing Leisure: Innovations in Happiness, Well-Being and Fun.* Helsinki: National Consumer Research Centre.

Picard, Rosalind. 1997. *Affective Computing.* Cambridge, MA: MIT Press.

Pike, Kenneth L. 1954. *Language in Relation to a Unified Theory of the Structure of Human Behavior.* Glendale, CA: De Gruyter Mouton.

Pine, B. Joseph, II, and James H. Gilmore. 1999. *The Experience Economy*. Boston: Harvard Business Review Press.

Polhemus, Ted. 1994. *Street Style*. London: Thames and Hudson.

Polson, Peter G., and Clayton H. Lewis. 1990. "Theory-Based Design for Easily Learned Interfaces." *Human-Computer Interaction* 5: 191–220.

Prahalad, C. K., and M. S. Krishnan. 2008. *The New Age of Innovation: Driving Co-Created Value through Global Networks*. New York: McGraw-Hill.

Priha, Päikki. 1991. *Pyhä kaunistus. Kirkkotekstiilit Suomen käsityön ystävien toiminnassa 1904–1950* [Sacred beautification: Church textiles and the work of the Friends of Finnish Handicrafts 1904–1950]. Helsinki: Taik.

Radice, Barbara. 1985. *Memphis: Research, Experiences, Results, Failures and Successes of New Design*. London: Thames and Hudson.

Rathgeb, Markus. 2006. *Otl Aicher*. London: Phaidon Press.

Redström, Johan. 2006. "Towards User Design? On the Shift from Object to User as the Subject of Design." *Design Studies* 27: 123–139.

Redström, Johan. 2017. *Making Design Theory*. Cambridge, MA: MIT Press.

Rhea, Darrel K. 1992. "A New Perspective on Design: Focusing on Customer Experience." *Design Management Journal* (Fall): 40–48.

Rizzo, Francesca. 2009. *Strategie di co-design: Teorie, metodi e strumenti per progettare con gli utenti* [Strategy of co-design: Theory, method, and structure of projects]. Milan: FrancoAngeli.

Robinson, W. S. 1950. "Ecological Correlations and the Behavior of Individuals." *American Sociological Review* 15: 351–357.

Rosenberg, Morris. 1990. "Reflexivity and Emotions." *Social Psychology Quarterly* 53: 3–12.

Ross, Philip. 2008. *Ethics and Aesthetics in Intelligent Product and System Design*. Eindhoven, Netherlands: Technishe Universiteit Eindhoven.

Sacks, Harvey, and Emanuel Schegloff. 1974. "A Simplest Systematics for the Organization of Turn-Taking for Conversation." *Language* 50: 696–735.

Säde, Simo. 2001. *Cardboard Mock-ups and Conversations: Studies in User-Centered Product Design*. Helsinki: UIAH.

Säde, Simo, Marko Nieminen, and Sirpa Riihiaho. 1998. "Testing Usability with 3D Paper Prototypes—Case Halton System." *Applied Ergonomics* 29: 63–73.

Salovaara, Juhani. 1985. *Kampaustyön ergonomia* [The ergonomics of hairdressing]. Helsinki: Taideteollisen korkeakoulun julkaisusarja A2.

Salvador, Tony, Genevieve Bell, and Ken Anderson. 1999. "Design Ethnography." *Design Management Journal* 10: 35–41. https://doi.org/10.1111/j.1948-7169.1999.tb00274.x.

Salvador, Tony, and Karen Howells. 1998. "Focus Troupe: Using Drama to Create Common Context for New Product Concept End-User Evaluations." In *Proceedings of Conference on Human Factors in Computing Systems CHI'98*, 251–252. New York: ACM.

Sanders, Elizabeth B.-N. 1994. "But Is It Useful? Testing beyond Usability." *Innovation* (Spring).

Sanders, Elizabeth B.-N. 2000. "Generative Tools for Codesigning." In *Collaborative Design*, edited by Stephen A. R. Scrivener, Linden J. Ball, and Andrew Woodcock, 3–14. London: Springer-Verlag.

Sanders, Elizabeth B.-N., and Uday Dandavate. 1999. "Design for Experience: New Tools." In *Proceedings of the First International Conference on Design and Emotion*, 87–92. https://zenodo.org/record/2631380/files/Design%20%26%20Emotion.pdf.

Schachter, Stanley, and Jerome E. Singer. 1962. "Cognitive, Social, and Physiological Determinants of Emotional State." *Psychological Review* 69: 379–399.

Schechner, Richard. 2006. *Performance Studies: An Introduction*. 2nd ed. New York: Routledge.

Schouwenberg, Louise, and Gert Staal. 2008. *House of Concepts: Design Academy Eindhoven*. Amsterdam: Frame Publishers.

Schultze, Gerhard. 1992. *Die Erlebnisgesellschaft*. Frankfurt am Main, Germany, and New York: Campus Verlag.

Scrivener, Stephen. 2000. "Reflection in and on Action and Practice in Creative-Production Doctoral Projects in Art and Design." *Working Papers in Art and Design* 1. herts.ac.uk/artdes/research/papers/wpades/vol1/scrivener2.html/.

Scrivener, Stephen. 2004. "The Practical Implications of Applying a Theory of Practice Based Research: A Case Study." *Working Papers in Art and Design* 3. https://www.herts.ac.uk/__data/assets/pdf_file/0019/12367/WPIAAD_vol3_scrivener_chapman.pdf.

Scrivener, Stephen A. R., Linden J. Ball, and Andrew Woodcock, eds. 2000. *Collaborative Design*. London: Springer-Verlag.

Segal, Leon D., and Jane Fulton Suri. 1997. "The Empathic Practitioner: Measurement and Interpretation of User Experience." In *Proceedings of the Human Factors and Ergonomics Society Annual Meeting* 41(1): 451–454. https://journals.sagepub.com/doi/10.1177/107118139704100199.

Sellen, Abigail, Phil Gossett, Richard Banks, Richard Harper, Sîan Lindley, Stuart Taylor, Tim Regan, et al. 2011. *Domesticating the Web*. Helsinki: Aalto.

Selznick, Philip. 1949. *TVA and the Grass Roots: A Study in the Sociology of Formal Organization*. Berkeley and Los Angeles: University of California Press.

Sevelius, Desirée. 1997. "Uusiotiiliä ja kierrätyslaattoja: jätemateriaalit rakennuskeramiikan raaka-aineina." Licentiate's thesis, UIAH, Helsinki.

Shedroff, Nathan. 1991. *Experience Design 1*. Indianapolis: New Riders Publishing.

Shott, Susan. 1979. "Emotion and Social Life: A Symbolic Interactionist Perspective." *American Journal of Sociology* 84: 1317–1334.

Shove, Elizabeth, Mika Pantzar, and Matt Watson. 2012. *The Dynamics of Social Practice: Everyday Life and How It Changes*. Los Angeles: SAGE.

Shove, Elizabeth, Matthew Watson, Martin Hand, and Jack Ingram. 2007. *The Design of Everyday Life*. New York: Berg.

Siikamäki, Raija. 2006. *Glass Can Be Recycled Forever: Utilisation of End-of-Life Cathode Ray Tube Glasses in Ceramic and Glass Industry*. Helsinki: UIAH.

Silverstone, Roger. 1994. *Television and Everyday Life*. London: Routledge.

Silverstone, Roger, and Eric Hirsch. 1992. *Consuming Technologies: Information and Communication Technologies and the Moral Economy of the Household*. London: Routledge.

Silverstone, Roger, Eric Hirsch, and David Morley. 2003. "Information and Communication Technologies and the Moral Economy of the Household." In *Consuming Technologies: Media and Information in Domestic Spaces*, edited by Roger Silverstone and Eric Hirsch, 15–31. London: Routledge.

Simon, Herbert A. 1969. *The Sciences of the Artificial*. Cambridge, MA: MIT Press.

Soini, Katja. 2015. *Facilitating Change: Towards Resident-Oriented Housing Modernization with Collaborative Design*. Helsinki: Aalto.

Soini, Katja, and Heidi Paavilainen. 2013. "Design Studio in the Field." In *Designing for Wellbeing*, edited by Turkka Keinonen, Kirsikka Vaajakallio, and Janos Honkonen, 87–98. Helsinki: Aalto.

Soini, Katja, and Antti Pirinen. 2005. "Workshops—Collaborative Arena for Generative Research." In *Proceedings of Designing Pleasurable Products and Interfaces DPPI'05*. Technische Universiteit Eindhoven, Netherlands. http://www2.uiah.fi/~kvirtane/julkaisut/KS-AP_Workshops_DPPI05.pdf.

Sorainen, Elina. 2006. *Aalto, ketju ja taatelitarha* [Wave, chain, and a date garden]. Helsinki: UIAH.

Spitz, René. 2002. *Hfg Ulm: The View behind the Foreground: The Political History of the Ulm School of Design*. Stuttgart: Edition Alex Menges.

Stappers, Pieter Jan. 2007. "Doing Design as a Part of Doing Research." In *Design Research Now*, edited by Ralf Michel, 81–91. Basel, Switzerland: Birkhäuser.

Sterling, Bruce. 2005. *Shaping Things*. Cambridge, MA: MIT Press.

Stock, Gregory. 1993. *Metahuman: Humans, Machines and the Birth of a Global Super-Organism*. London: Bantma Press.

Stryker, Sheldon. 1980. *Symbolic Interactionism: A Social Structural Version*. Menlo Park, CA: Benjamin/Cummings.

Stryker, Sheldon. 1986. "The Vitalization of Symbolic Interactionism." *Social Psychology Quarterly* 50: 83–94.

Suchman, Lucy. 1987. *Plans and Situated Actions: The Problem of Human Machine Communication*. Cambridge, UK: Cambridge University Press.

Summala, Heikki, Juha Karola, Alessandro Couyoumdjian, and Igor Radun. 2003. *Head-On Crashes on Main Road Network: Trends and Causation*. Helsinki: Finnish Road Administration.

Summatavet, Kärt. 2005. *Folk Tradition and Artistic Inspiration: A Woman's Life in Traditional Estonian Jewelry and Crafts as Told by Anne and Roosi*. Helsinki: University of Art and Design.

Swann, Cal. 2002. "Action Research and the Practice of Design." *Design Issues* 18: 49–61.

Tennenhouse, David. 2000. "Proactive Computing." *Communications of the ACM* 43: 43–50.

Thackara, John. 2019. "Bioregioning: Pathways to Urban-Rural Reconnection." *She-Ji* 5: 15–28.

Tharp, Bruce M., and Stephanie M. Tharp. 2018. *Discursive Design: Critical, Speculative, and Alternative Things*. Cambridge, MA: MIT Press.

Thomas, William I., and Dorothy S. Thomas. 1928. *The Child in America*. New York: Knopf.

Thomas, William I., and Florian W. Znaniecki. 1918–1920. *The Polish Peasant in Europe and America*. Chicago: University of Chicago Press and Boston: Gorham Press.

Tilley, Alvin R., and Henry Dreyfuss. 2002. *The Measure of Man*. New York: John Wiley & Sons.

Toffler, Alvin. 1980. *The Third Wave*. New York: Morrow.

Turkle, Sherry. 1995. *Life on the Screen: Identity in the Age of the Internet*. Cambridge, MA: MIT Press.

Vaajakallio, Kirsikka. 2012. *Design Games as a Tool, a Mindset and a Structure*. Helsinki: Aalto.

Vaajakallio, Kirsikka, and Tuuli Mattelmäki. 2014. "Design Games in Codesign: As a Tool, a Mindset and a Structure." *Codesign: International Journal of Cocreation in Design and the Arts* 10: 63–77.

Valle-Noronha, Julia. 2019. *Becoming with Clothes: Activating Wearer-Worn Engagements through Design*. Helsinki: Aalto.

van Kollenburg, Janne, and Sander Bogers 2018. *Data-Enabled Design*. Eindhoven, Netherlands: Technishe Universiteit Eindhoven.

Vejlgaard, Henrik. 2008. *Anatomy of a Trend*. New York: McGraw-Hill.

Verganti, Roberto. 2009. *Design-Driven Innovation: Changing the Rules of Competition by Radically Innovating What Things Mean*. Cambridge, MA: Harvard University Press.

Vezzoli, Carlo, Cindy Kohtala, Amrit Srinivasan, Liu Xin, Moi Fusakul, Deepta Sateesh, and J. C. Diehl. 2014. *Product-Service System Design for Sustainability*. Sheffield, UK: Greenleaf Publishing.

Vihma, Susann. 1995. *Products as Representations: A Semiotic and Aesthetic Study of Design Products*. Helsinki: UIAH.

Vihma, Susann, ed. 2010. *Design Semiotics in Use*. Helsinki: Aalto.

Viña, Sandra, and Tuuli Mattelmäki. 2010. "Spicing Up Public Journeys—Storytelling as a Design Strategy." *Linköping Electronic Conference Proceedings* 60(7): 77–86. https://ep.liu.se/konferensartikel.aspx?series=ecp&issue=60&Article_No=7.

Virtanen, Katja. 2005. *Asukaslähtöisen perusparantamisen kehitystarpeet: IKE-esitutkimus* [Development needs for resident-oriented building renovation and modernization: IKE pilot study]. Helsinki: Ministry of the Environment.

Voûte, Ena, Pieter Jan Stappers, Elisa Giaccardi, Sylvia Mooij, and Annemiek van Boeijen. 2022. "Innovating a Large Design Education Program at a University of Technology." *She Ji* 6: 50–66.

Wakkary, Ron. 2021. *Things We Could Design: For More than Human-Centered Worlds*. Cambridge, MA: MIT Press.

Wasson, Christina. 2000. "Ethnography in the Field of Design." *Human Organization* 59: 377–388.

WDO. 2022. "Industrial Design Definition History." https://wdo.org/about/definition/industrial-design-definition-history/.

Weiser, Mark. 1991. "The Computer for the 21st Century." *Scientific American* 265(3): 94–104.

Wensveen, Stephan. 2004. *A Tangibility Approach to Affective Interaction*. Delft, Netherlands: Delft University of Technology.

Wensveen, Stephan, and Ben Matthews. 2015. "Prototypes and Prototyping in Design Research." In *Routledge Companion to Design Research*, edited by Paul. A. Rodgers and Joyce Yee, 262–276. London: Routledge.

Winograd, Terry, and Fernando Flores. 1987. *Understanding Computers and Cognition: A New Foundation for Design*. Reading, MA: Addison-Wesley.

Wittgenstein, Ludwig. 2009. *Philosophical Investigations*. 4th ed. Malden, MA: Wiley-Blackwell.

Wu, Yiying. 2017. *Bicycles and Plants: Designing for Conviviality and Meaningful Social Relations through Collaborative Services*. Helsinki: Aalto.

Wu, Yiying, and Ilpo Koskinen. 2021. "Plant Hotels: Designing the Imaginary Foundations of Communities." *CoDesign*. doi: 10.1080/15710882.2021.1991958.

Wu, Yiying, Jack Whalen, and Ilpo Koskinen. 2015. "Nothing Makes Sense: New Aesthetics of Experience in Self-Organizing Services." In *Empowering Users through Design*, edited by David Bihanic, 249–266. Dordrecht, Netherlands: Springer.

Ylirisku, Salu. 2013. *Frame It Simple! Towards a Theory of Conceptual Designing*. Helsinki: Aalto.

Ylirisku, Salu, and Jacob Buur. 2007. *Designing with Video: Focusing the User-Centred Design Process*. Dordrecht, Netherlands: Springer.

Zimmerman, John, Jodi Forlizzi, and Shelley Evenson. 2007. "Research through Design as a Method for Interaction Design Research in HCI." In *Proceedings of Conference on Human Factors in Computing Systems CHI'07*, 493–502. New York: ACM.

Index